OBSERVATIONS

ON THE PAST AND PRESENT STATE

OF

FIRE-ARMS,

AND ON THE PROBABLE EFFECTS IN WAR

OF

THE NEW MUSKET:

WITH

A PROPOSITION FOR REORGANIZING THE ROYAL
REGIMENT OF ARTILLERY
BY A SUBDIVISION INTO BATTALIONS
IN EACH SPECIAL ARM OF

Garrison, Field, and Horse Artillery,

WITH

SUGGESTIONS FOR PROMOTING ITS EFFICIENCY.

By COLONEL CHESNEY, D.S.L. & F.R.S.

The Naval & Military Press Ltd

❖

Reproduced by kind permission of the Central Library,
Royal Military Academy, Sandhurst

Published by
The Naval & Military Press Ltd
Unit 10, Ridgewood Industrial Park,
Uckfield, East Sussex,
TN22 5QE England
Tel: +44 (0) 1825 749494
Fax: +44 (0) 1825 765701
www.naval-military-press.com

TO

FIELD MARSHAL H.R.H. PRINCE ALBERT,

K.G., K.T., K.P., G.C.B., AND G.E.M.G.;

COLONEL AND CAPTAIN-GENERAL OF THE HONOURABLE ARTILLERY COMPANY;

PRINCE CONSORT OF HER MAJESTY.

I GRATEFULLY avail myself of the permission so condescendingly given by your Royal Highness, to dedicate the following observations on the state and prospects of the Royal Regiment of Artillery to your Royal Highness. I do this the more gladly and hopefully since the name of the Consort of our Most Gracious Sovereign is, in the present day, the harbinger of progress; and consequently will be that of brightening hope to a branch of the Military Service which so urgently requires that its present imperfect state should become a matter of history.

The pages now submitted to your Royal Highness contain explanations of, rather than discussions upon, those points in my former publication which appear to have been misunderstood by the Committee of the House of Commons, with the addition of some historical notices, calculated to show the tardy progress of our Artillery, compared with that of other nations.

Page 122, and those which immediately follow, contain the serious fact, that the proportion of Artillery taken into the field in late years has been diminished; and this fact might have been made even more prominent by going still farther back. The despatches of Marlborough show that, both as regards the number of guns and their employment, Artillery held then a more important place than at this moment, when the Sovereign, the Leader of her Army, and the Nation itself, so far eclipse even the brilliant period of Queen Anne.

Three defects, among others, in the existing state of the Artillery Corps will, it is presumed, be made palpable in the pages now humbly offered for the consideration of Your Royal Highness.

The first is, that the British Artillery does not bear a due proportion to the rest of the Army.

The second, that the Officers belonging to this

Service are worn out before they attain the rank of Colonel-en-Second. And

The third, that the separation of the Artillery branches of the Ordnance from the rest of the Service, is fraught with the most serious evils to the Army at large.

If the nation which stands highest in the scale of civilization has lately benefited by the centralization of genius, talent, and industry in her capital, may not her Artillery Service also gain by the study of this branch, as organized in other nations, and her officers take a higher place, at this critical period, when their peculiar arm is, in consequence of recent improvements, threatened with the rivalry in the field of another, to which it has hitherto been considered superior in importance?

With the greatest respect,
I have the honour to subscribe myself,
Your Royal Highness'
Most obedient and most humble Servant,

F. R. Chesney,
Colonel Royal Artillery.

Packolet, Kilkeel, Ireland,
December 20, 1851.

CONTENTS.

CHAPTER IX.

CHAPTER X.

CHAPTER XI.

CHAPTER XII.

OBSERVATIONS,

&c. &c.

CHAPTER I.

ON ACQUIRING THE ART OF WAR BY STUDY.

Reconstruction of the Royal Regiment of Artillery proposed in 1849.—Objects to be attained by this change.—Evidence given before the Committee of the House of Commons.—Combination of Artillery with other Arms.—Basis on which this question should be considered.—Theory and experience compared.—Observations by Plutarch, Folard, Puységur, and others, on the Study of War.—A knowledge of War may be theoretically acquired.—The celebrated Campaign of Turenne against Montecuculi in 1675.—Result of their Manœuvres.—Montecuculi is opposed to Condé.—Tactics of Gustavus Adolphus and those of Frederic the Great as described by Guibert and Jomini.—Battle of Leuthen illustrates the Tactics of the latter.—General Houchard and change of Tactics by the French Directory.—The Emperor Napoleon pursues the new system.—Naval Tactics propounded by Clerk of Eldin.—System of breaking the Enemy's Line.—Rodney defeats the Count de Grasse by a similar manœuvre.—Other examples of the advantages derived from Study without Experience.

Reconstruction of the Royal Artillery proposed in 1849.

UNDER the impression that a judicious reconstruction of the Artillery would gradually remedy its well-known defective state of promotion, and at the same time confer other important advantages on this branch of the service, the writer, without attempting to go deeply into this important subject, ventured to submit his ideas to the authorities in 1849; and his pamphlet so far attracted attention, that he was summoned from Ballincollig by the Select Committee of the House of Commons on the Ordnance and Army Expenditure, in May of that year. But however ready the noble Chairman (Lord Seymour) and most of the Members were to give due attention to a project which had the good of the Service for its object, the author's reception in some instances was almost calculated to remind him of the words of our immortal bard:—

"To thee I call, but with no friendly voice."

Those who are sufficiently interested in the subject to take the trouble, will find the evidence given during this peculiar and unexpected examination, from page 368 to page 384 of the Blue Book, and that which immediately followed and appeared to be consequent upon it, from page 384 to page 396 of the Second Report of the Army and Ordnance Expenditure, ordered by the House of Commons to be printed, 12th July, 1849.

Evidence given in consequence before a Committee of the House of Commons.

It must be evident that in a pamphlet dedicated to the Deputy Adjutant-General of the Royal Artillery himself, there could not have been the expectation, and certainly there was not the least intention on the writer's part, that anything should have been said calculated to give offence to the authorities. It was, in fact, little more than the outline of a plan by which, or by something similar, those who have the ability as well as the *power*, might at the present expense create a more efficient Service than that which exists; and even the attempt to accomplish this object, from whomsoever it may come, claims the most patient consideration.

We are told that self-defence is the first law of nature common to all created beings; but trusting that the Committee were well aware that the writer's object was simply such a reconstruction of the Artillery as might eventually exchange the state of age and decrepitude which prevails amongst the senior officers of the corps, for one of such efficiency both of mind and of body as will permit them to serve the country usefully, perhaps the most suitable answer to all remarks and objections will be to endeavour to explain more fully and more clearly those passages, which from being much condensed have been misunderstood, and even occasionally their accuracy questioned.

Being one of the many officers to whose lot garrison instead of field service fell during the war, the writer, it may seem, is not warranted by practical experience* in touching upon the

* "He has had no experience in the field, and therefore I presume is not very competent to give his opinion against those who have."—No. 6004, Evidence of Major-General Sir H. D. Ross, K.C.B., Second Report Army and Ordnance Expenditure, ordered by the House of Commons to be printed, July 12, 1849.

important subjects now to be considered. In his case, however, it may be observed that service in the field, had it been his good fortune, would only have been performed with the grade of Lieutenant, which he held up to the close of the war in 1815; and if he is not mistaken, there is no officer of the Artillery alive who had a more extensive command in the late war than that of a troop of horse Artillery, or a company attached to field guns, which, as regards the application of the arm itself to the purposes of war, was but a section of a section. It is not so much the management of one of these batteries, however perfect it may have been, as the general adaptation of the Artillery service to, and its suitable combination with, the other two arms, that is to be considered. Principles, not details, therefore, should regulate the organization of an army as well as of its component parts, having due regard to the principles adopted by the most experienced nations, deriving at the same time useful experience from the modern battles of greatest note, and the causes, which may and will occur again, of their being lost and won.

The proportion of artillery to be regulated by general principles.

With reference to this question, it has been maintained that theory, apart from actual warfare, is far preferable to the latter when deprived of the benefits derived from the former.

Without, however, going to this length, and

far from depreciating the advantages of active service, which are much more envied than underrated, the preceding observations have been made, in order that the present discussion may take broad and general, instead of restricted grounds; and if military writers of celebrity have correctly stated that study alone has been sufficient to enable individuals of moderate acquirements, some of them being civilians, to master the principles and science of war, it is hoped that the writer will not be considered presumptuous in venturing to touch upon one of its branches, to which some attention has been given during a lengthened period of service.

A knowledge of war may be gained by study alone.

In the first book of Livy, and in Plutarch's life of Marcus Brutus, it will be seen how much may be accomplished without the practical part of war. This, says Folard, in his chapter on the importance of reading to a commander, is to ordinary men a science, to which nothing contributes more than the study of good books. Marshal Puységur is still more explicit, and thus deals with his subject:*—

The opinions given by Puységur and other military writers,

"Ainsi loin d'être persuadé qu'il faille attendre que l'on fasse la guerre pour apprendre comment on la doit faire, je crois au contraire que les plus grands capitaines qui ne se sont formés que par la pratique seule, ont été sujets à faire bien des fautes, dont ils se seraient garantis s'ils avoient étudié les

* Art de la guerre par principes et par règles par M. le Maréchal de Puységur, tome i., pp. 2, 3.

règles et les principes des différentes parties de la guerre; c'est ce que je prouverai par ce qu'ils en ont écrit eux-mêmes.

that a knowledge of tactics, such as that of fortification,

"J'entreprends donc de faire voir que sans guerre, sans troupes, sans armée, et sans être obligé de sortir de chez soi, par l'étude seule, avec un peu de géometrie et de géographie, on peut apprendre toute la théorie de la guerre de campagne depuis les plus petites parties jusqu'aux plus grandes, et cela en la même manière que le Maréchal de Vauban par la théorie renfermée dans les livres qu'il nous a laissés, et par la pratique qu'il a établie en conformité, nous apprend l'art de fortifier, d'attaquer, et défendre les places, ce qui même journellement est enseigné par des personnes qui n'ont jamais été à la guerre, ni fait travailler à fortifier les places."

Elsewhere he says:—

may be acquired by means of study.

"Que nous devons être persuadés que les ordres de bataille se peuvent apprendre comme on apprend les fortifications, l'attaque et la défense des places, et qu'il en est de même de toutes les autres parties dont il ne nomme que les titres; parties qui regardent les operations de l'esprit, et qui sont les plus grandes et les plus essentielles."*

Strategic operations replaced by simpler tactics.

It is well known that Puységur's opinion was formed after having had the advantage of studying the strategic operations of the most talented commanders of ancient and modern times, among which those of the great Turenne against Count Montecuculi were the most instructive. It is not however intended, in touching upon this and other campaigns, to go into the subject of military tactics beyond such brief notices as may suffice to

* Art de la guerre par principes et par règles par M. le Maréchal de Puysegur, tome i., pp. 23, 24.

carry the reader cursorily through the effete but complicated system of warfare, so carefully and systematically pursued in the eighteenth century. This was the memorable turning point in the early part of the present age, when French civilians devised a new system of warfare, which, as far as the writer is aware, was previously unknown, and consequently was not the result of former experience.

In the campaign of 1675, so justly considered a masterpiece of the skill of the old school, the eyes of Europe were intently fixed on the two talented leaders above mentioned, on whose skill and judgment the result of the war between the King of France and the Emperor of Germany was to depend. They were nearly of the same age, with equal advantages in point of instruction, and having become Captains in the art of war by means of study, their operations were necessarily regulated by prescribed rules which left little or nothing to fortune; consequently, neither could reasonably expect to gain a victory through the fault of his enemy.

The campaign of 1675 given as a specimen of the old school.

On the 21st May of that year, the campaign was commenced by Turenne, who endeavoured to anticipate the passage of the Rhine by his adversary near Strasburg, while the latter made a feint of besieging Philipsburg, hoping to draw the Viscount from his position, and thus open a passage into Alsace. Suspecting the real object of this movement, Turenne, whilst threatening Montecuculi with part of his force, so placed the bulk of his army that he could at pleasure operate towards either of the bridges across the Rhine in his rear, and accordingly, though each point was frequently menaced in turn, the enemy failed in his attempt. To lessen the difficulty of maintaining a line of about three leagues with only 20,000 men, Turenne removed one of the bridges nearer to the other, thus reducing his line to two leagues; and this position he maintained for several weeks against the army of his adversary. The latter, hoping to cut off or exhaust the supplies of the French army, threw himself on one of its flanks along the Rhine, but was speedily menaced by the watchful Turenne, with the view of causing him either to fight or retire. A series of manœuvres followed in the more open country, till, at the close of a campaign of 66 days, the warrior game of chess, which had been played with such consummate skill and perseverance on the gigantic board of the plains of Baden, seemed to have reached the point when a checkmate promised to terminate its course.

The manœuvres of Turenne and Montecuculi at its close.

After both Generals had exhausted all the finesses of art to starve, intercept, surprise, or at least gain some advantage before engaging in battle, Turenne at length succeeded in enclosing the left wing of the Imperialists, whose retreat was thus partially cut off. He had scarcely exclaimed, "It is done—I have them—they cannot escape me any more, and I shall reap the fruit of so fatiguing a campaign," when a chance cannon-shot ended his career, and no one remained who could appreciate or was prepared to take advantage of the fruits of his patience and consummate skill.*

Death of Turenne and close of this memorable campaign.

Montecuculi was subsequently opposed to the great Condé, in a campaign which proved to be the last conducted by this warrior, to whose system of tactics succeeded that of Gustavus Adolphus and the improvements of the Prussians under Frederic the Great. In the protracted warfare of this great tactician, two things were prominent,—the introduction of a rapidly-moving artillery, and the general use of columns of infantry selected for manœuvres during attacks, which were carefully covered and supported by the former arm.

With a skilful sovereign to lead, and the resources of the kingdom at command, the improvement of the army, especially when opposed to numerous forces led by talented generals, could not be otherwise than rapid. The battles of this great captain differ from those of his predecessors

Frederic the Great's manœuvres, and.

* From the History of Henri De la Tour d'Auvergne, Viscomte de Turenne, Marshal-General of France, vol. i., p. 462 to 489. James Bettenham, London, 1785.

by his troops being so completely in hand, that they could move, even when close to the enemy, with a degree of rapidity hitherto unknown. A Prussian battalion was compared by the warrior himself to a moveable battery, the impetus of whose charge tripled its power and made one man equal to three. This facility in manœuvring enabled Frederic at pleasure either to turn the enemy, or suddenly to make an attack elsewhere on discovering some weak point of the enemy's position, for which reason he usually assumed the offensive, and almost invariably sought instead of waited for the enemy. The tactics resulting from the seven years' war, which will be found detailed in Guibert's "Traité de Tactique," and again in that of Jomini, "Traité de grandes opérations Militaires," may be summarily described as the employment of masses of artillery so as to cover the columns of attack, the basis of which was, but in appearance only, a greater display of troops on the feigned than on the real point of attack; the success, as a matter of course, mainly depending upon success in deceiving the enemy. Frederic, for instance, at the battle of Leuthen, made demonstrations against the right wing of Marshal Daun, who at first suspected the feint; but being at length deceived by renewed and continuous demonstrations, he sent all his reserves to strengthen that wing. The king observing that

tactics of the Seven Years' War.

Frederic's tactics in the battle of Leuthen.

his feint was at length successful, proceeded to execute an oblique attack against the left wing of the Austrians, which was in consequence overpowered, and a complete victory gained by the Prussians.

But lest the preceding instances should seem to belong to practical rather than to scientific warfare, others may be cited in which experience could have had no share: the system of the present century is an instance in point. Instead of following or improving upon the systems already known, our neighbours devised another of a novel but effective kind, almost at the commencement of the revolutionary war. General Houchard not having followed up by an immediate attack the advantages gained by him over the British force at Hoondscoote, he was thus addressed by Barrière:—

A new system of tactics adopted by the French Directory

"Instead of accumulating the troops in large masses on particular points, you have divided them into separate detachments, and the generals entrusted with their command have generally had to combat superior forces. The Committee of Public Safety, fully aware of the danger, had sent the most positive instructions to the generals to fight in large masses; you have disregarded their orders, and in consequence reverses have followed."*

This aptly illustrates a system of warfare,

* Le malheureux général fût condamné à mort pour avoir su vaincre, mais pour n'avoir pas su profiter de la victoire. Histoire critique et militaire des campagnes de la Révolution, &c., par le Général Jomini, tome v., p. 349. Paris, 1811.

which, as far as the writer is aware, was laid down for the first time by the Committee of Public Safety; and this theory, which was propounded by civilians and followed out by the republican generals, became the mainspring of brilliant successes, even of those of Napoleon himself; the action being always preceded, however, in his case, by the use of an overpowering fire of artillery.

and followed by Napoleon.

It may be remembered that at a somewhat earlier period a no less remarkable change had been devised, also by a civilian, in another branch of tactics, which comes more peculiarly home to Great Britain and her dearest interests.

Naval tactics improved by Clerk of Eldin, and

It appears that John Clerk, of Eldin, was the inventor of one of the most important parts of the modern British system of naval attack, namely, breaking, or rather cutting the enemy's line, so as to double upon and overpower one portion of it.

The uninterrupted success of single ships, and even of small squadrons of British men-of-war, compared with the undecisive results attending the encounters of large fleets with those of the enemy during the greater part of the American and preceding wars, led Mr. Clerk to conclude that the French must have adopted a new system of tactics, by which our fleets, when acting on a large scale, had been hitherto baffled. Finding that the French fleets, when extended in line of battle, invariably courted a leeward position, and that they too frequently succeeded in raking and even disabling our fleet in coming down, end on, to the attack, Mr. Clerk perceived how this disadvantage might be avoided. The mode by which an attack can be safely made on an enemy's fleet, whether to windward or leeward, will be readily understood; and it can, in fact, be mathematically demonstrated, that the chief strength of the fleet may be brought to act against the weakest and most vulnerable part of the opposing line.

In the one case, instead of bearing directly down, a fleet to windward being arranged in three or more divisions as the service may require, should continue the pursuit of an enemy, like a single ship, in nearly a parallel line of approach, and should moreover confine the attack to as many ships as can be reached or cut off in the centre or rear. For even if the enemy's fleet have the superiority in sailing, the swiftest of those vessels which, in the pursuit, are to windward, will necessarily outsail and intercept the dullest sailers to leeward, so that the enemy must either abandon these to their fate, or return to bring on a general and close engagement under disadvantageous circumstances.

Application of his theory.

In the other case, that of a chase to windward, the courses of both fleets being in angular lines, the distance would continue the same if the speed and other advantages on both sides were equal.

But as any accident to the masts or rigging of a single ship must delay the whole fleet, unless indeed she were to be allowed to fall to leeward into the hands of the enemy, it may be concluded that unavoidable accidents will, in the course of a few days' pursuit, enable the fleet to leeward to *fetch* and cut off some part of the enemy's line, which in this case must either be abandoned to its fate, or, as before, the leading ships must return to bring on a close engagement with much disadvantage.

Rodney uses a similar system against Count de Grasse in 1782.

The experiment was tried for the first time with brilliant success by Rodney against Count de Grasse, on the 12th of April, 1782, in consequence, as has been asserted by Mr. Clerk's relations and friends, of his having indirectly communicated his project to Sir George Rodney. This important claim has not, however, been established; and it has, on the contrary, been successfully contended that the chance position of the two fleets and Sir George's skill, at a critical moment, gave rise to the operation in question, without any previous knowledge of the ideas developed by Mr. Clerk in 1779. But even in this case, it is impossible to deny to the latter the efficacy of the method, or the honour of having first demonstrated the advantages of a system which was made known to the world by him; a system which, in consequence of the victory of

Lord Howe on the 1st of June, 1793, caused the French convention to pass a decree of death against that captain who should suffer the line to be cut.*

Since Mr. Clerk was not bred to the sea, and had never performed even a single voyage, it could not have been from practical experience that he was led to make discoveries in naval tactics, which had escaped the observation of professional men; nor, with reference to less important matters, could it have been from personal knowledge that the naval and military equipments proposed by the author to be furnished by Government for the Euphrates expedition, proved to be well adapted for carrying out both by land and water, through distant regions, the various objects to be accomplished.

Other instances bearing on the same point.

But two other instances which have particular interest for the corps of Artillery, may be noticed as better illustrating the ideas of Marshal Puységur; the one having reference to the efficient improvement of the Service, the other to the management of the arm itself during a campaign. A word will probably suffice to remind the reader

* See Playfair's Memoirs of Mr. Clerk. Also Edinburgh Review, vol. vi., p. 301; and an Essay on Naval Tactics, systematical and historical, with explanatory plates, in four parts, by John Clerk, Esq., of Eldin, Fellow of the Scottish Society of Antiquaries (first printed in 1782); compared with article in the Quarterly Review, vol. lxii., pp. 50—80.

how much has been done by the late General Millar towards the accomplishment of the former object, both by land and sea, without the inventor, as far as the writer is aware, having shared in any campaign.

The other instance regards the career of the most distinguished of all Artillery officers, the late Sir Alexander Dickson. It is understood that a plan so simple as to be contained in a sheet of foolscap paper was submitted by this officer to the Duke of Wellington regarding the Artillery service, at the time when preparations were being made for the sieges of Cuidad Rodrigo and Badajos. Although Captain Dickson had served previously, particularly at Monte Video, and the blockade of Malta, a project connected with extensive field operations could not have been so much the result of personal experience, as of the knowledge derived from those military studies, which, as the writer well remembers, occupied this officer's spare time at Portsmouth, and which, doubtless, laid the foundation of the high place in military history now occupied by the brilliant and well-earned fame of Major-General Sir Alexander Dickson, G.C.B.

CHAPTER II.

ARTILLERY OF EARLY TIMES.

Derivation of the name of Artillery.—The Balista, Catapulta, &c., the earliest kind of Artillery.—Power of Machine Artillery.—Used at Jerusalem 1000 years B.C.—Their invention attributed to the Sicilians, 3000 B.C.—Artillery of our Men-of-war in the Thirteenth Century.—Artillery a name of general application to all Engines of War in former times.—Early use of Gunpowder in China.—Combustible Materials and Explosive Instruments of the Chinese.—Guns used in China A.D. 85.—Stone Shot and Fire Machines used in Caïfong-fou in the Thirteenth Century.—Guns were known in China long before the time of the Missionaries.—Ancient breech-loading Jinjals and other Guns still found in China; also Gunpowder.—Ancient Bells found at Pekin.—The art of Founding long known in China.—Gunpowder known in India at the time of Alexander's Invasion.—Inflammable Oils, Explosive Substances, &c., of the Indians.—Fire-arms mentioned in the Gentoo Laws.—Their early use in India.—Extensive use of Artillery by the King of Delhi, A.D. 1258; Muhammed Shah, A.D. 1368; and Mahmud Shah, A.D. 1482.—Artillery at Sea and on Land found by the Portuguese in the East in 1498; also Fire-ships.—Ancient Guns at Boorhampore, and Colossal Bombard found at Moorshedabad.—Indian breech-loading Guns, similar to Venetian Pateraroes, at Woolwich.—Early preparation and use of Gunpowder and Propellant Machines by the Arabs.—Introduced by the Saracens into Europe.—Greek Fire.—Stone Cylinder and Gunpowder used by the Arabs at the Siege of Alexandria.—Artillery used by the Moors in Spain.

Proposed sketch of artillery.

BEFORE entering upon the intended sketch of the field artillery of the principal armies of Europe, it will be proper to give some account of the arm in general, of which it forms an essential part.

Reference to the past will show that the march of artillery towards its present efficiency has been exceedingly slow, compared with the time that the basis of this formidable power, namely, the means of propulsion, have been known to the world.

and derivation of the name.

Originally the name appears to have been Arcualia, from Arcus, a bow,* and it appears to have included all sorts of missiles, as well as the engines by which they were propelled. The common sling, which is still retained by the Arabs on the banks of the Upper Euphrates, was in all probability the first kind of artillery, and the arrow, one of the succeeding stages of improvement, which in the sequel embraced a variety of machines, such as the balista, catapulta, espringal, trebuchet, mangonel, and others which it would be tiresome to mention. They, may, however, be considered generally either under the head of balista, or catapulta; the former hurling stones, varying between 2 lbs. and 330 lbs., and even occasionally 500 lbs., to a distance of 90 yards,† and the latter propelling arrows and iron bolts some 200 yards. Both kinds of machine may be traced back to 1,000 years before our era, when Uzziah had engines in Jerusalem, "invented by "cunning men, to be upon the towers and upon "the bulwarks, to shoot arms and great stones "withal." ‡

Power of machine artillery,

Plutarch's account makes the origin of such instruments later; they were, he says, inventions

* Vossius, de Vitiis Sermonis, lib. iii., c. i.

† Etudes sur le passé et l'avenir de l'Artillerie, par le Prince Louis Napoleon Bonaparte, tome ii., pp. 46, 47, 48, compared with Grose's Military Antiquities, vol. i., pp. 357-364.

‡ 2 Chron., c. xxvi., v. 15.

of the Syrians; and Diodorus, as well as Plutarch, attributes the invention, which was in reality a readoption, to the Sicilians, about 300 B. C., when the battering-ram, which is much older, also reappeared.* That the name under consideration was applied to the preceding and other machines in the fourth century is evident from Vegetius, who calls balistæ, onagri, scorpiones, arcubalistæ, fustibuli, and fundæ, engines of artillery.† So recently as the thirteenth century, the artillery of our men-of-war consisted of stones and darts discharged from machines, which also threw pots of fire, quicklime, and other combustible materials. Even as late as the fourteenth century it is stated by Froissart that there were collected at Yprès, A. D. 1384, two tons *of artillery*, chiefly arrows, which were shot into the town.‡ This circumstance would seem to account for some of the mistakes into which historians have occasionally fallen in their accounts of battles, in consequence of the term artillery having been still applied to the arrow after that great change had commenced which, as the result of the invention of gunpowder, eventually restricted the use of this name almost entirely to large ordnance, such as cannons, mortars, howitzers, and in the present day, rockets.

and application of the name in ancient times.

Did circumstances permit, it would be far from

* Diod. Sic., lib. xiv., p. 91; and Plutarch's Apothegms.

† Lib. iv., c. xxii.

‡ Vol. vi., p. 298, translation by Johnes. London, 1806.

an unpleasing task to attempt to do something more than give the following brief notice of gunpowder and its application to the service of mankind, which there is reason to believe first took place in the East.

Early use of gunpowder in China.

Beginning with the Chinese, probably the oldest portion of the human family, and whose government is, undoubtedly, the most ancient in the world, Sir George Staunton observes, that "Nitre is the natural and daily produce of China and India, and there, accordingly, the knowledge of gunpowder seems to be coeval with that of the most distant historic events. Among the Chinese it has been applied at all times to useful purposes, such as blasting rocks and removing great obstructions, and to those of amusement in making a vast variety of fireworks. It was also used as a defence, by undermining the probable passage of an enemy, and blowing him up. But its force had not been directed through strong metallic tubes, as it was by Europeans soon after they had discovered that composition." Yet this invention did not prove so decisive for those who first availed themselves of it as to mark distinctly in history the precise time when its practice first took place.*

There is some little difficulty in distinguishing

* Embassy from the King of Great Britain to the Emperor of China, by Sir George Staunton, Bart. Second edition: Bulmer and Co., London, 1798, vol. ii., pp. 292, 293.

between inflammable substances and those which in a more advanced state were used as propellants; but it is remarkable, in connexion with the former, that China snow, and China salt, are names given to saltpetre by Arabian writers, and white Chinese fire, and red Chinese fire, show, as the names express, that pyrotechny belongs to this ancient people.*

Other inflammable compositions have also retained their Chinese names, as Joung-ko (hive of bees), another kind of weapon no less terrible than the Ty-lai (terrestial thunder), and belonging to the same age; Leho-yas (devouring fire), Le-ho-toung (box or tube of fire), Le-tien-ho-kien, that is to say, a globe containing the fire of heaven, the effects of which appear to have been similar to those of the Greek fire, and to have been known several centuries before the Christian era.† Combustible materials.

A more formidable instrument, the thunder of the earth, is thus described. A hollow globe of iron, large enough to contain a bushel of gunpowder, with which it was charged, mixed with fragments of iron and brass. One or more of these being placed where the enemy was expected to advance, it was exploded at the proper moment Explosive instruments, and

* Bibliographical Index to the Historians of Muhammedan India, by H. M. Elliott, Esq. (Calcutta, 1849, pp. 345, 346), compared with Du feu Grégois, et des Origines de la Poudre, par M. Reinaud et M. Favé, p. 176. Paris, 8vo, 1845.

† Du feu Grégois et des Origines de la Poudre, p. 178.

to cause destruction.* Another instrument, called a fire tube, was thus prepared. A round bamboo about 2½ feet in circumference was chosen, and lashed with hempen cords to prevent it from splitting, and a strong wooden handle was added to each tube, making the length of the whole 5 feet. It was then charged with several layers of powder of different kinds, over which were placed fire-balls, made of a certain kind of composition, which, being thrown to a distance of 100 feet by the discharge, consumed any materials they might encounter.† Such was the so-called impetuous dart of the Chinese, which, on being ignited, projected a violent flame from the tube with a great noise against the enemy.

Fire-tubes of the Chinese.

Early use of gunpowder and fire-arms in the East.

In addition to these primitive fire-arms, a later writer has in a great measure established the important fact that both gunpowder and metal fire-arms were known and used in China long before the Christian era. According to the report of M. Paravey,‡ it is mentioned in Chinese books that in 618 B. C., during the Taing-Off dynasty, they used a cannon bearing this inscription—"I hurl death to the traitor, and extermination to the rebel;" and from a circumstance mentioned by Captain Parish, who accompanied Lord Ma-

* Du feu Grégois et des Origines de la Poudre, p. 183.
† Ibid.
‡ To the Academie des Sciences, 1850.

cartney to China, it may be inferred "that the claim of the Chinese to a very early knowledge of the effects of gunpowder is not without foundation." "The soles of the embrasures," he observed, "were pierced with small holes, similar to those used in Europe for the reception of the swivels of wall-pieces. The holes appear to be part of the original construction of the wall, and it seems difficult to assign to them any other purpose than that of resistance to the recoil of fire-arms."* This would carry back the use of jingals to more than three centuries B. C., since the barrier of the great wall was finished about 221 years before the Christian era.†

Jinjals used in China 221 B.C.

Ufano says,‡—

"Le Rd. père Fr. Andrieux d'Aquirre, Provincial de l'ordre de St. Augustin, Isles Philippines, dans sa rélation sur le royaume de la Chine dit qu'en l'an de nostre Seigneur 85, ceste invention eut son commencement en ses quartiers, et qu'en auculnes provinces maritimes du dit royaume on trouvait encore pour le jourd'huy quelques pièces d'artillerie fort anciennes de telle façon et proportion, tant de fer que de cuivre, avec mémoire de l'année de leur fonte, et engraveure du nom, des armes blason du Roi Vitey, qui en fut l'inven-

Guns used in China A.D. 85.

* Embassy to China, by Sir George Staunton, Bart. Bulmer and Co., London, 1798, vol. ii., p. 198.

† The General History of China, by P. Duhalde, vol. i., p. 29, and vol. ii., p. 76. J. Watts, London, 1741.

‡ Artillerie, c'est-à-dire Vraye Instruction de l'Artillerie de toutes ses appartenances. En espagnol par Diego Ufano, Capt. d'Artillerie au Château d'Anvers, traduit en français par Jean Théodore de Bry, Bourgeois d'Oppenheim; c. i., p. 1. Folio, Francfort, 1614.

teur, et qu'on sçait par monuments des histoires anciennes et véritables, que le dit roi, grand nigromentien et enchanteur, après avoir coniuré ses enchantements, le malin esprit qui lui en montrast la façon et l'usage, fut le premier qui en usa contre les Tartares au royaume de Pégu, et en la conquête des Indes Orientales. Le mesme, est racompté par plusieurs Portugais qui ont navigué et costoyé ces quartiers: comme aussi le père Herruada et ses compagnons. Et s'accorde fort bien avec une lettre du Capitaine Artrède au Roy, nostre Sire, l'avisant et lui racomptant en grande diligence toutes les particularités de ce grand royaume; disant qu'en tous ces quartiers là on use de mesmes armes et l'artillerie comme par de-ça, et que dès longtemps ou y trouve quelques vieilles piedrières mal faites: qui les fontes modernes sont de meilleure façon et estoffe que celles de par de-ça, et beaucoup plus fortes et durables; qu'en chascune ville il y a Arsenal, auquel entre autres choses, ou prepare la poudre et fond l'artillerie."

Father Amyot states, on the authority of a Chinese manuscript which he caused to be translated, that all the effects of gunpowder were incontestably known in China many centuries before the Christian era, and that from the very commencement of this epoch, artillery had been greatly developed.*

Called terrestrial thunder, A.D. 200.

According to the same author, terrestrial thunder (ty-lai) was successfully employed about A. D. 200 by Koung-ming, who was not, however, the inventor of such means of destroying an enemy;† and the use of cannons, or rather of stone mortars, is more particularly mentioned

* Mémoires des Missionaires de Pekin, tome viii., p. 331, et suivants.

† Du feu Grégois et des Origines de la Poudre, p. 177.

A. D. 757, when the general Li-Kouang-pi, of Thang's army, constructed guns to throw stones of 12 lbs. weight a distance of 300 paces.* But, as Captain Favé observes, even if the statements of Father Amyot should be exaggerated, it is beyond all doubt that the Chinese made use of artillery in the thirteenth century. For when besieged in Caïfong-fou by the Mongols in 1232, they threw round stone shot of different weights against their enemies. They also had in this town, as additional means of defence, fire-machines, ho-pao, peacocks,† called Tchin-tien-lei, which being filled with powder and ignited, burst with a noise like a clap of thunder, that might be heard more than 100 ly, the effect of which extended to the distance of half an acre all round the spot where it exploded.‡ The preceding accounts appear to be so well authenticated as to leave little or no doubt that some kind of artillery, as well as an inflammable composition like the Greek fire, were in use at the period in question.

Use of stone shot and explosive instruments by Chinese in the 13th century.

With respect to the former, Du Halde's account of guns having been cast by the missionaries for the Chinese, has given rise to the erroneous im-

Guns in China before the time of the missionaries.

* Du feu Grégois et des Origines de la Poudre, p. 186.

† Mailla observes that the Chinese still use the word *pao* to designate a cannon.

‡ Mailla, Histoire Générale de la Chine, tome ix., p. 166; Gaubil, Histoire des Mongons, 1739, 4to, pp. 34–37; Histoire des Mongols, par M. H. Quatremère, tome i., pp. 135, 136, notes; and A. Danduli, Chronicon. ap. Muratori, xii., 448.

pression that the use of artillery was only made known to this people about A. D. 1636; whereas it is manifest from another passage given by this writer, that only an improved system of casting was then introduced among the Chinese, who had, on the contrary, been long in possession of rudely-constructed cannon.

Ancient bombards of Nankin,

"Though the use of gunpowder is very ancient in China, artillery is but modern, and they have seldom made use of powder since it was invented but for fireworks, in which the Chinese excel: there were, however, three or four bombards at the gates of Nankin, ancient enough to make one judge that they had some knowledge of artillery, and yet they seemed ignorant of its use, for they serve for nothing but to be shown as curiosities: they have also pateraroes in their buildings on the sea-coast, but have not skill enough to make use of them."*

and other parts of China.

Antique jinjals loading at the breech by means of moveable chambers, as well as pateraroes, probably similar to those noticed by Du Halde in 1621, and possibly, also, like those noticed above as existing at the remote period of king Vitey, may still be seen in some of the northern parts of the empire, and give, considering the unchanged character of the people, a claim to priority in the use of these rude engines. The existence of seven ancient bells at Pekin, each weighing 620,000 lbs., is also sufficient proof that the Chinese have long been in possession of the art of

* The General History of China, by P. Du Halde, vol. ii., pp. 78, 79. London, J. Watts, 1741.

founding, the process of which is so complicated and so difficult.*

Whilst the knowledge of gunpowder and its propellant application have been claimed for the Chinese, equal, if not still greater antiquity in both, has been supposed to belong to another Asiatic nation, viz., the Hindús, whose knowledge, owing to their greater intercourse at a remote period with other nations, was more likely to spread than that of the self-excluding people of the celestial empire.

Gunpowder or some other propellant compound,

That some kind of formidable missile, capable of striking an enemy at a distance, was known to the Indians almost from the earliest historic period is evident, since it is alleged that if Alexander the Great had even succeeded in passing the Hyphasis, he never could have mastered the strongholds of the sages. For, says the historian, if an enemy were to make war upon them, he would be driven off by means of tempests and thunders, as if sent down from heaven. Such was the case when they were attacked by the Egyptian Hercules and Bacchus, on which occasions the sages remained, as it were, unconcerned spectators till an assault was attempted, when it was repulsed by whirlwinds and thunders, hurling destruction on the invaders.† Another author,

known in India during Alexander's invasion.

* Anquetil's Summary of Universal History, vol. v., p. 552.

† Philostrati Vit. Apollon., lib. ii., c. xxxiii.

Themistius, also mentions that the Brahmins fight at a distance by means of lightnings and thunders.* In his supposed letter to Aristotle, Alexander mentions the terrific flashes of flame showered on his army in the burning plains of India.†

Ctesias says that the people living on the banks of the river Indus, manufactured a kind of oil, which being enclosed in earthen jars, and thrown against wood-work, caused a flame so strong, that it could only be extinguished by throwing a quantity of mud upon it. It was manufactured solely for the king, and no one else was allowed to have it.‡ Ælian, quoting the preceding author, says, the oil had such strength that it burnt not only wood, but men and animals, or anything else that it touched, and that the Indian kings took the cities by its means; that no battering-ram, or other poliorcetic machine, could resist its flames. Earthen jars filled with this mixture being thrown against the city gates, and thus broken by the concussion, the burning oil necessarily spread, and since it could not be extinguished by any ordinary means, it served the purposes both of offensive weapons and of fighting men.§

Great power of the Indian combustibles.

* Orat., xxvii., p. 337, ap. Dutens, Origine des découvertes attribuées aux modernes, p. 196; and Maurice's History of Hindustan, vol. i., p. 144.

† Dante, Inferno, canto xiv., 31.

‡ Ctesias, India Excerpta, xxvii. Ed. Baer. p. 356.

§ De Naturâ Animal., lib. v., c. iii.

Elsewhere it is said that the Indian philosophers, on perceiving that the troops of Alexander were alarmed at the approach of Fúr, with some 2,000 elephants, suggested the use of an iron horse, with the figure of a man, also in iron, which, being filled with naphtha and placed on a carriage, might explode on coming in contact with these animals. Accordingly, Alexander, having collected all the blacksmiths and artisans, constructed 1,000 of these machines, which, being fired at the proper moment, destroyed and burnt many of Fúr's elephants, and his army fled in confusion.* The use of a naphtha arrow is also mentioned in the Sháhnámah, which has an additional interest from the geographical position of Persia, which caused her to be a link between India and western nations.

Fire-arms and other propellants mentioned in the Gentoo laws.

In the code of Gentoo laws,† the following remarkable sentence occurs:—"The magistrate shall not make war with any deceitful machine, or with poisoned weapons, or with cannon and guns, or any kind of fire-arms." Gunpowder has been known in Hindustán and China beyond all periods of investigation. In Sanscrit it bears the name of "Aigmaster," weapon of fire. It was at first used as a kind of dart or arrow, tipped with fire, and discharged from a bamboo. One pro-

* Sháhnámah, Turner Mason's edition, vol. iii., p. 1108.
† Halked's edition, iii.

perty was that, during flight, it separated into several streams of flame, each of which took effect, and could not be extinguished. But, he adds,* this kind of aigmaster is now lost. A cannon is called "Shataghnee," or the weapon that kills 100 men at once; and it was invented by Viscarme, the Vulcan of the Hindús.†

Early use of fire-arms and explosive substances in India.

In a work which goes into this subject at some length, it is observed, with reference to the preceding and other circumstances, that we may conclude that some kinds of fire-arms were known and used in the early stages of Indian history; that the missiles were explosive; that the time and mode of ignition were dependent on pleasure; that certain kinds of projectiles were made to adhere to gates, buildings, and machines, which they set on fire from a considerable distance; that in all probability saltpetre, the principal ingredient of gunpowder, and the cause of its detonation, entered into the composition, particularly as Gangetic India is richly impregnated with this material; and another ingredient, sulphur, abounds towards the north-western side of the territory.‡

Sulphur and saltpetre in India.

It is moreover clear, from their medical works, that the Indians were acquainted with the constituents of gunpowder, and possessed them in

* Halked's edition, iii. † Ibid.

‡ Bibliographical Index to the Historians of Muhammedan India, pp. 373, 374.

great abundance, though we have no positive account of the invention itself.* Professor Wilson, in his lecture on this subject, says that the writings of the Indians make frequent mention of arms of fire and of rockets, which appear to be of Indian invention, though not expressly mentioned as such in Sanscrit writings; but they had been long used when the European armies came in contact with them.†

Artillery earlier amongst Eastern than Western nations.

Passing, however, from the earliest use of incendiary projectiles by the Hindús, in which some kind of great gun appears to have been included, the use of the latter, as well as that of propellant gunpowder, seems to have been pretty general amongst this people antecedently to their adoption by western nations. Chased, the Hindú bard, says:—"Oh! chief of Gajné, buckle on your armour, and prepare your fire machines;" and he adds (stanza 257), that the culivers and cannons made a loud report when they were fired off, and the noise of the ball was heard at the distance of about 10 coss, or nearly 1,445 yards. As this took place about A.D. 1200, during the Ghorian dynasty, the fact of cannon balls having been propelled by means of gunpowder in India, at that early period, appears to be established;‡

* Professor Wilson's Lecture, in the Athenæum, July 8, 1848.

† Ibid.

‡ Bibliographical Index to the Historians of Muhammedan India, pp. 350, 351.

while, as far as has been traced in these pages, the use of artillery is not even mentioned by any European writer before the fourteenth century, at which period masses of this arm formed part of the Indian field equipment. The writer whose painstaking researches have thrown so much light on this subject, mentions that in A.D. 1258 the wazir of the king of Delhi went out * to meet the ambassadors sent by Halákú, the grandson of Genghis Khan, with 3,000 carriages of fireworks.† A.D. 1368, Muhammed Sháh Bahmiani took, among other spoil, 300 gun-carriages, and from these being so numerous, it may be inferred, even if there were no other corroborating evidence, that cannon had been in use for some time.

Extensive use of artillery, A.D. 1368, and of shells, A.D. 1380.

Passing over other instances mentioned by Mr. Elliott as having occurred in the interim, it appears that Mahmud Sháh, of Guzerat, embarked gunners in the fleet which he despatched against the pirates of Bulsor, A.D. 1482.‡ Two years earlier he breached the walls of Champanir, and fired shells into the palace of the rajah. These instruments of destruction, also, were early used by the Muhammedan Indians. The Portuguese, on their arrival in the East in 1498,

* Bibliographical Index to the Historians of Muhammedan India, p. 351.

† Ibid., p. 353.

‡ Ibid.

found artillery much in use. A culiver, carried by one of the Nagres, was fired at intervals when Vasco de Gama entered Calicut during that year; and only two years later, A.D. 1500, the Portuguese vessels were attacked by a Hindú called Zamori, who had 3,000 Nagres, 300 of whom were archers, and 40 of the number were armed with matchlocks, and had likewise several frames provided with ordnance.* At this period a Guzerat vessel fired several guns at the Portuguese invaders. We find fire-vessels in use amongst the Indians of Calicut in 1502, and in the following year the Zamorin's fleet was armed with no less than 180 guns, and fire-ships were also used by them successfully.†

Artillery both at sea and on land,

In the following year, Albuquerque was opposed by a fleet of 80 Indian ships, carrying 380 guns. The Hindús on this occasion had fire-ships, or floating castles, raised on a double boat; and it is also stated that the Múhammedans used grenadoes and other fireworks.‡

In 1511, the people of Malacca not only opposed the Portuguese with cannon, but also defended their streets with mines of gunpowder;

when the Portuguese reached India.

* Bibliographical Index to the Historians of Múhammedan India, by H. M. Elliott, Esq., Calcutta, 1249, p. 355; and p. 503, vol. iii., of the History of the Rise of Múhammedan Power in India, by John Briggs, M.R.A.S., Lieut.-Col. in the Madras Army. Longman, Rees, Orme, and Co., London, 1829.

† Ibid., p. 505; Faria-e-Souza, tome i., part i., c. vii.

‡ Ibid., Lieut.-Col. Briggs, pp. 505, 507.

and at sea they used floats of wild fire. Muhammed, king of Java, is said to have brought 3,000 guns to bear, out of 8,000 which he had at command.*

Artillery used in Java,

A Portuguese prisoner of the fleet of Sequeira states that among those taken "were many of great size, and one gun of great beauty, which the king of Calicut had lately sent to the king of Malacca as a present." †

In the attack on Malacca, the Portuguese manned a junk, or large native vessel, with guns, in order to batter a bridge which was one of the main defences of Malacca. De Barros ‡ has the following passage on this subject:—

"As soon as the junk had passed the sand-bank, and had come to an anchor, a short way from the bridge, the Moorish artillery began firing at her. Some guns discharged leaden balls at intervals, which passed through both sides of the vessel, doing much execution among the crew. In the heat of the action, Antonio d'Abreu (the commander) was struck in the cheek from a fusil (espingardão), carrying off the greater number of his teeth."§

and muskets at Malacca.

The town of Malacca extended for a league

* History of the Rise of the Muhammedan Power in India, by Lieut.-Col. Briggs, vol. iii., p. 510, and Faria-e-Souza, tome i., part ii., c. vii.

† Decades of Jao de Barros, decade ii., book vi.

‡ Decade ii., book vi., c. v.

§ O qual junco tanto que passou o banco d'arêa, a foi surto hum pedaço da ponte, começou a artilheria dos Mouros descarregar nelle; alguma da qual lançava pelouro de chumbo do tamanho de hum tiro de epera, que passaoa ambos os costados do junco, furendo minto damno na gente; na qual furia de fogo com hum espingardão foi Antonio d'Abreu ferido pelas queixadas, levando-lhe a maior parte dos dentes.

The Portuguese attack Malacca.

along the shore, and had no other fortifications than some stockades, and armed vessels anchored along the shore. Alphonso Albuquerque attacked it, in 1511, with 800 Portuguese and 200 natives of Malabar, armed with swords and shields. He was beaten off in his first attack, and forced to retire to his ships, but succeeded in the second.

Castagnedor, another Portuguese historian of India, says of the resistance made by the Malays of Malacca—

"Surely never, from the commencement of our conquest of India until this day, was an enterprise undertaken by us so arduous as the attack of the bridge, nor one in which so much artillery was employed (by the enemy), nor so many men employed in the defence. From the play of the enemy's artillery also we suffered much loss before effecting the landing."*

Artillery was likewise used in other parts of the east at this period. In 1521, Pigafetta, the companion and secretary of Magellan, and narrator of his voyage round the world, states that the walls of the town of Borneo, which he had visited, were mounted with six pieces of iron ordnance and fifty-six pieces of brass.†

Ancient guns of India and China.

There is reason to believe that the gun in use at this remote period had a moveable breech, like the oldest description of jinjal, which the

* Castagnedor, vol. iii., p. 190.
† Primo Viaggio intorno al Floto Terraqueo. Milano, 1800.

Guns in Northern China.

writer has seen in the northern part of the unchanged empire of China, and also in some of the forts towards the north-western side of India. When the writer passed through Boorhampoor, in 1837, several ancient pieces of ordnance, taken, as was understood, in the neighbouring fort of Asseergur, were then being broken up, for the purpose of selling the metal. In some of these pieces the bore was formed of bars of iron welded together, round which the brass, forming the rest of the gun, appears to have been cast. Others were constructed entirely of iron bars welded together, and secured by means of iron rings driven on outside the gun. One of the latter kind of pieces was understood to weigh about 18 tons, and one of the former from 10 to 12 tons.

Thinking that the subject might not be undeserving of attention, the author ventured to write to the Military Secretary at Bombay, expressing a hope that two of each kind of these singular guns, might be reserved as specimens of early Indian artillery.

Colossal bombard found at Moorshedabad;

A breech-loading gun of a still ruder construction has been recently discovered in the bed of the Bagretti river, at Moorshedabad, in Bengal. According to the description and sketch given in the "Illustrated News" of the 18th October last, the bore of this colossal bombard is 18½ inches in diameter, and its length 12 feet 2 inches, independently of the moveable chamber, which is 4 feet 2 inches. The latter, which is of the same construction as the rest of the piece, fits into the former when loaded, and seems to have been secured by firmly lashing a set of rings on each portion to one another, no doubt with the additional support of a block of wood, to prevent the breech from separating by the force of the explosion. Like the pieces recovered from the "Mary Rose," the cylinder, as well as the breech, of this enormous piece is formed of massive longitudinal bars of wrought iron, encircled by eleven powerful rings, encircling it at 11 inches apart. The bore is, however, uneven, being of very rude workmanship. But it should be borne in mind that such pieces were intended for stone shot, to be fired with small charges of powder, which, imperfect as is their construction, they would doubtless have borne. dimensions of this great gun.

Small guns of this description appear to have been used for the defence of such tunnelled entrances as those of Chittoor, and the tubes themselves being built into the masonry, the moveable parts (for there were, in some instances, more than one to each gun) were brought in succession, ready for use. These instruments are similar to the Venetian pateraroes, in the Repository at Woolwich, which will presently be described. The Smaller pieces of ordnance.

Ancient jinjals in China.

date is the chief difficulty with respect to those in India; but as regards the remoter part of the east, the holes, which, as already shown from Sir George Staunton's work, are to be found in the great wall of China for similar instruments, seem to carry their use back at least to the third century before Christ; and, as a matter of course, the ruder bamboo tube, already noticed, must have been much earlier.

If, as may be presumed from the preceding and other corroborative evidence, cannons were an eastern invention, the knowledge of their use might have reached Europe either through the Arabs or the Venetians, at a later period.

Preparation and use of gunpowder

With respect to the earliest of these sources, a passage in one of the Arabic works in the Escurial collection, about 1249, undoubtedly describes gunpowder;* and another Arabian writer describes the use of cannon in the years 1312 and 1323.† From these and other circumstances bearing upon this subject, there seems, says a celebrated historian of these days, little reason to doubt that gunpowder was introduced through the means of the Saracens into Europe.‡ Indeed, there is every reason to believe that the various receipts for the mixture of sulphur, saltpetre, and

by the Arabs.

* Caseri, Bibl. Arab. Hispan., tome ii., p. 7.

† Ibid.

‡ View of the State of Europe during the Middle Ages, by Henry Hallam, Esq., p. 361. John Murray, 1818.

charcoal which have been found amongst the Arabs, from time to time, were obtained by this people from the east, possibly as far back as the ninth century of our era, when not only commercial intercourse existed between China and the Arabs through the Persian Gulf,* but also by the land expeditions which were sent from Arabia about the period in question, to make conquests in the east. In connexion with this subject, it appears that the manjanik,† a propelling machine, said to have been copied from the Persians, was used by the Arabs A. H. 9, when Muhammed besieged Taif. On another occasion, Hajjaj said to Muhammed Kásim, "Fix the manjanik, shorten its foot, and place it on the east; call the master of the manjanik, and tell him to aim at the flag-staff." So he brought down the flag-staff, and it was broken, at which the infidels were sore afflicted. This event is said to have occurred A. H. 93, that is, A. D. 711 or 712.‡ This instrument had been previously used A.D. 200, by Jazymah the Second, and its invention was ascribed by the Arabs to the devil.§

Propellant machines

long known by the Arabs.

The fire-playing machines, probably such as

* Expedition to the Rivers Euphrates and Tigris, by Lieut.-Col. Chesney, R.A. Longmans, 1849, vol. ii., pp. 161, 571, 572, 577, 578.

† Either from mangannum or machina.

‡ Bibliographical Index, &c., by H. M. Elliott, Esq., p. 346.

§ Ibid., p. 347.

Fire-playing machines.

the manjanik, were used at the capture of Alore, shortly after that of Daibal. On another occasion, it is stated that the Arabs took vessels filled with fireworks, and threw them on the seat, or howdar, fixed on the backs of the elephants, which caused the affrighted animals to become unmanageable, and to run off, breaking the ranks of the Hindús. This was the first time that incendiary fire like that of the Greeks, appears to have been used by the Arabs, and not long afterwards it became common in Europe. When Krummis, prince of Bulgaria, took the town of Mesembria, in Thrace, he found thirty-six fire-tubes (siphones), with a provision of liquid fire. The Greek emperor, Leo, introduced hand-tubes for the same purpose, that is, for the use of Grecian fire, between 890 and 911; and in the eleventh century, fire-pieces of iron and other metal are mentioned as having been on board the ship of the emperor Alexis I., and in those of the king of Tunis. Similar traces are found of the use of a composition like the Grecian fire in the thirteenth century, at which period gunpowder, as such, replaced it in Europe.*

Fire-tubes used in Europe with an inflammable composition, A.D. 890.

Friar Bacon, who had, it is supposed, partly derived his knowledge from an ancient manu-

* Allgemeine Deütsche Real-Encyclopædia, or Conversation's Lexicon, 8vo edition. Leipzig. 8vo, from 1833 to 1837, vol. ix., pp. 747, 748, art. Schiesspulver.

script, was aware of the component parts of gunpowder,* which are, in fact, detailed in his Treatise de Nullitate Magiæ, Oxford, 1216.

Bacon's knowledge of gunpowder.

Whether this was borrowed knowledge, or simply the result of his own ingenuity, is uncertain, but gunpowder became generally known through Bartholdus Schwartz, in 1320; and as regards its practical use at a remote period, Arabian authors state that it was first employed by being discharged from cylinders excavated in a rock, during one of the earliest sieges of Alexandria.†

Gunpowder discharged by the Arabs from a stone tube.

At a comparatively recent period, namely in 1771, an experiment was tried successfully by discharging stones from a similar kind of mortar, which may still be seen in a part of the rock of Gibraltar. This excavation is 4 feet long and 36 inches diameter at the muzzle, and it resembles the bore of a mortar, carefully polished. A tompion being placed over the charge, and some 15 cwt. of stones over this, and being without a vent or touchhole, it was discharged by means of a hollow reed and copper tube, which, being passed through the centre of the stones and tompion, ignited the powder near its extremity.

* Marcus Gracchus, in his Liber Ignium ad Combusendos Hostes, gives as the proper ingredients a mixture of 6 lbs. of saltpetre, 2 lbs. of brimstone, and 1 lb. of charcoal.

† Expedition to the Rivers Euphrates and Tigris, by Lieut.-Col. Chesney, vol. ii., pp. 499, 500.

The discharge was formidable, the stones spreading right and left for a distance, in some cases, of 500 yards.*

It is not easy to determine how long the Arabs had been acquainted with the use of gunpowder; but it appears that artillery, as distinguished from incendiary compositions, was used by this people in the early part of the twelfth century.

Artillery used by the Moors between A.D. 1118 and A.D. 1249.

Condé, in his History of the Moors in Spain, says that artillery was used by the Moors against Zaragossa, in 1118, and that a culverin of 4-lb. calibre, named Salamonica, was made in 1132. In 1157 the Spaniards took Niebla, in which the Moors defended themselves valiantly by machines which threw darts and stones by means of fire.† The Moorish king, Abd'almumen, attacked and captured Mohadia, a fortified city near Bona, and took it from the Sicilians, by the same means, in 1156.

In the year 1249, the Egyptian Al Malik Alsaleh describes a military instrument in which the Arabs burnt powder, which made a horrid noise, destroying and burning all things to ashes.

A mortar employed by the Arabs, 1391.

In 1280, artillery was used against Cordova, after which period this arm is frequently mentioned. The Arab Al Mailla, in his History of the Saracens, describes a species of mortar, by

* Military Antiquities respecting a History of the English Army, by Francis Grose, Esq., vol. ii., pp. 168–171. Stockdale, London.

† Universal History, vol. xiv., p. 305.

means of which, with the use of powder and fire, houses were reduced to ashes, A.H. 690, or A.D. 1291. In 1306, or 1308, Ferdinand IV. took Gibraltar from the Moors by means of artillery.

Abdallah Ibn El Khatib, in his History of Spain, says that large machines were used with powder to throw burning globes into the city of Baza, in 1312 and 1323. Martos was attacked in the same manner in 1326, Alicante in 1331, and Tariffa in 1340.*

Also shells against Baza, A.D. 1312;

Ibn Nason ben Bia, of Grenada, mentions that guns were adopted from the Moors, and used in Spain, in the twelfth century, and that balls of iron were thrown by means of fire in 1331.

and iron shot, A.D. 1331.

* Allgemeine Deütsche Real-Encyclopædia, or Conversation's Lexicon, vol. ix., pp. 747, 748, art. Schiesspulver.

CHAPTER III.

PROGRESS OF ARTILLERY IN EUROPE.

Crakeys of war used by Edward the Third.—Artillery at the Sieges of Cambray and Quesnoy.—Used at Algesiras, 1343.—Cannon invented at Bruges, A D. 1346.—Artillery at Cressy doubtful.—Extensively used in the Fourteenth and Fifteenth Centuries.—Its progress under Charles VII. of France.—Construction of the earliest great Guns.—Italian Artillery of the Fourteenth Century, and British of the Fifteenth Century.—English Guns of the Sixteenth Century.—Guns from the Mary Rose, &c.—French Artillery of the Fourteenth Century. —Mode of serving the Guns.—First Train of Siege Artillery.—Waggon Batteries.—Guns in the Musée de l'Artillerie at Paris and the Repository at Woolwich.— Light Guns made of Copper, Rope, and Leather, used by Gustavus Adolphus.—Portable Guns in Flanders and Scotland. —Artillery used by the Burgundians.—The earliest Field Artillery was small.—Great improvement in Artillery under Charles VIII. of France.—Proportion of Artillery to the Army.—Further improvement under Francis I.—Improvement in Small Arms.—Their first use at the Battle of Pavia.—Light Artillery at Cerisolles.—War of the Protestant League.—Use of the Prolonge by the French.—Progress of Artillery under Henry IV. of France, Maurice of Nassau, and Gustavus Adolphus.—Battle of Leipzig.—Passage of the Lech.—Battle of Lutzen.—Death of Gustavus Adolphus.—Defective state of the Imperial, and advanced state of the Swedish, Artillery.

Guns employed against Cambray, A.D. 1339, and Quesnoy, 1340.

WITH the exception of certain weapons called "crakeys of war," which Edward III. had during his campaign against the Scots in 1327,* the use of artillery in Europe appears to have been confined to Spain till about the year 1339, when ten

* The Life and Acts of the Most Victorious Conqueror Robert Bruce, King of Scotland; by John Barbour, Archdeacon of Aberdeen, carefully corrected from the edition of 1620. Edinburgh, 1758, 4to.

cannons were prepared for the siege of Cambray by the noble Chevalier Cardaillac.*

Artillery at Quesnoy.

When the Duke of Normandy besieged Quesnoy in the following year, 1340, the town being well provided with men at arms and heavy artillery, the people *allowed* the besiegers to hear their cannon, which flung large iron bolts in such a manner as caused the French to retreat.†

When Alphonsus XI., king of Castille, besieged Algesiras in 1343, the Moorish garrison threw among their enemies certain thunders (ballistas del tueno) through long mortars or troughs of iron.‡ Three years later, the Consuls of Bruges authorized an experiment to be made by a tinman of their city. The cannon was of iron, and had a square instead of a cylindrical bore, from which a cubical iron shot of about 11 lbs. weight was discharged, and passed through both walls of the town; but as it also killed a man, the experiment was not repeated.§

A cannon with a square bore fired at Bruges, A.D. 1346.

It has been stated that Edward III. owed his great victory at Cressy in the same year, 1346, to the effect produced by some pieces of artillery placed in front of his army. This important cir-

Artillery at Cressy, doubtful.

* Etudes sur le passé et l'avenir de l'Artillerie, par le Prince Louis Napoleon Bonaparte, Preface, tome i. p. vii.

† Ibid., tome ii., p. 75; and Froissart's Chronicles, vol. i., p. 190. Longman, Hurst, Rees, Brown, and Co., 1805.

‡ Military Antiquities, &c., by Francis Grose, Esq., p. 377.

§ Mémoires sur la Poudre à Canon, de M. Léon Lacabane; in the Mémoires de l'Academie de Bruxelles; and in the Archives Philosophiques et Littéraires, tome ii. Gand.

Artillery used at Calais in 1347;

cumstance appears to rest chiefly on the authority of Villani,* who is not supported by Froissart or other chroniclers, for, although they enumerate Edward's forces in much detail, they do not mention the guns. The latter, namely, espringals, bombards, and crossbows, are however mentioned as having been placed by Edward on the tower which he constructed for the siege of Calais during the following year;† but the writer has not succeeded in finding any mention of artillery at the subsequent battle of Poictiers, nor, indeed, any sufficient proof that guns were brought into the *field* by the English about this period. Cannons, bombards, and Grecian fire were used by the Prince of Wales to reduce the castle of Romozantin in 1356.‡ Two years later, the numerous artillery of St. Valery did great execution amongst its besiegers.§ But this arm was used on a still more extensive scale when Richard II. attacked St. Malo in 1378, for he had full 400 cannons which fired day and night against the town and castle, but without success. Oudenarde also resisted successfully, the citizens having covered their houses with earth, to prevent them from being set on fire by the artillery of the Gantois and

extensively against St. Malo, in 1378, and

* G. Villani, lib. xii., c. lxvii. His words are, in describing the effect of Edward's cannon, "Colpi delle bombardi;" adding, it was as if God thundered, "con grande uccisione di genti e sfondamento di cavalli."

† Froissart, tome i., p. 243; lib. i. c. cccxv.

‡ Froissart's Chronicles, vol. ii., p. 300.

§ Froissart, lib. i. c. cxix.

Flemings.* Eight years later, 1386, the English ships took two French vessels, on board of which were divers great guns and engines for beating down walls, also a great quantity of powder that was more worth than all the rest.† From the commencement of the fifteenth to the middle of the sixteenth century, the use of artillery is mentioned in various sieges, for defence as well as attack, as at Bourges, Etampes, and in the same year, or 1420, at Melun; at Meaux, in 1422; Orleans, in 1428; Zurich, 1444; and Constantinople, in 1453.‡ The besieging batteries consisted of bombards both of great and small calibre, the latter being intended to keep up a fire, so as to prevent the besieged from repairing the mischief done by the previous discharge, during the interval consumed in reloading.§ In order to throw the projectiles over the walls into a town, the guns were pointed at a high angle, so as to give more power to the missile; and this continued to be the system of firing so long as large stone shot were used.‖ Towards the middle of the century, trenches and approaches were introduced as a means of attack, and instead of trusting to a kind

In other sieges subsequently.

Stone shot fired at a high elevation.

* Froissart, lib. lvii., p. 80.

† Hollinshed's Chronicles of England and Ireland, vol. ii., p. 777. Longman, Hurst, Rees, and Co., London, 1808.

‡ Etudes sur le passé et l'avenir de l'Artillerie, par le Prince Louis Napoleon Bonaparte, tome ii., pp. 77–98.

§ Ibid., tome ii., p. 63.

‖ Ibid., p. 67.

Direct fire supersedes the use

of wooden screen or parapet as a protection, guns were covered by barrels filled with earth. Soon afterwards breaching batteries were constructed by the two brothers Bureau, who, by this judicious change, demonstrated the irresistible power of artillery against walls, and enabled Charles VII. of France to recover in the short space of a year and ten days, all the places which had been previously taken by the English.* Bombards and other heavy artillery, which had for some time been employed, in addition to dart and stone throwing machines, in reducing towns and castles, were now beginning to supersede the latter; but although a certain description of light guns had existed about a century, they were seldom taken into the field. Whilst in a rude state, the difficulty of moving even the smallest pieces, confined the latter weapons almost exclusively to sieges. Indeed, the time unavoidably consumed in reloading, gave the archer a manifest advantage in the field, which he continued to retain till a later period, when those improvements had taken place in artillery and hand guns, which will now be briefly mentioned.

of ancient machine artillery.

1st. With regard to what is known of the invention of heavy artillery. Passing over the rock-formed mortar, already mentioned, of the

* Etudes sur le passé et l'avenir de l'Artillerie, par le Prince Louis Napoleon Bonaparte, tome ii., p. 99.

Arabs, and another made in the ground by the Poles at a later period, to hurl stones against the enemy in a similar manner.* The first great guns were usually formed of iron bars fitted together lengthways, hooped with iron rings at intervals, and generally wider at the mouth than at the chamber.

Specimens of ancient guns.

Amongst the specimens of early artillery to be seen in the Repository at Woolwich, there is a Venetian paterara of wrought iron, something like the description just given. A strong cylinder of rude workmanship forms the body of the piece, which is strengthened by hoops of iron placed at intervals. The diameter of the bore is 2½ inches, and its length 2 feet 5 inches. It was loaded by means of a moveable breech or chamber 7½ inches in length, which being charged and placed in the bore, was secured by a wedge of iron in rear of this moveable chamber. As this piece is similar to some that have been brought from China, it may be presumed that the mode of its construction was acquired from the Chinese; probably during the early intercourse which took place with the Celestial Empire through the Black Sea, towards the end of the fourteenth century.†

The Venetian paterara

like the Chinese guns.

Whether derived from Venice or some other source, the use of artillery appears to have been

Italian artillery of the 14th century,

* Military Antiquities, &c., by Francis Grose, Esq., p. 388.

† Expedition to the Euphrates and Tigris, by Lieut.-Col. Chesney, R.A , F.R.S., &c., vol. ii., pp. 585, 586.

Italian Artillery of the 14th century,

common in Italy during the fourteenth century; for there are a great many specimens of ancient artillery of iron as well as of brass, belonging to this period in the Hotel de Ville at Bologna. Eleven pieces of ordnance are enumerated in a catalogue prepared in the year 1381, when they were already ancient: one, which is of copper, carried a ball of 361 lbs. weight; another, also of copper, is cylindrical; and three others, which are of iron, are of a square form, possibly similar to that already mentioned as having being fired at Bruges in 1346.

and British of the 15th century.

Five wrought-iron bombards are preserved in the Musée de l'Artillerie at Paris, which were, it is said, abandoned by the English at the town of Meaux in 1422.

There are two guns belonging to this period in the Repository at Woolwich, which, from being similar to others in the Tower, are marked "pattern" of Henry VI. A.D. 1426. Both are of bars of iron, strengthened by rings hammered round them at intervals. The longer, which is in two pieces, is 7 feet 5 inches long, the shorter, 3 feet 6 inches, having in each case a bore of 4·2 inches diameter. A portion of the bore of one of these pieces has a brass cylinder, into which a moveable breech is fitted. The latter is of brass, and resembles one of the old wooden tompions. The breech of the other gun has been lost.

Guns of the time of Henry VI.

The Repository at Woolwich also contains other guns made of bars of iron, hooped and welded together. They are in a more advanced state, and belonging to a period which has been approximatively ascertained. By the successful and ingenious exertions of Mr. Dean, several guns of brass and iron have been recovered from a man-of-war called the "Mary Rose," which owing to the weight of her armament, sunk on the coast of France in the year 1545 with a crew of about 600 men, who perished on this occasion. One of the brass guns which carried a stone shot of 4·5 inches diameter, appears to have been cast hollow. One of the iron guns which was recovered at the same time is much ruder; and some idea of its construction may be formed by imagining an iron cylinder of 8·5 inches diameter, and 8 feet 7 inches long, formed by means of an overlap throughout its length, round which a succession of iron hoops, or rather rings of 3 inches square, appear to have been driven on whilst red hot. The gun is secured in its proper place by means of wedges and a block of wood, let into the carriage in the rear of the gun. The carriage itself is nothing more than the section of a large tree, partially squared with a hatchet, and having a groove cut about four-fifths of the length of the piece, and half its diameter, into which the gun is sunk and secured by bolts, in the

English guns in the 16th century,

recovered from the "Mary Rose."

Dimensions of the iron guns,

their carriages.

same way as the barrel of a musket is fastened into the stock, and it is kept in its place by means of a block of wood let into the carriage, against which the breech of the piece rests. It has no trucks or any kind of gear, so that it appears to have been kept by its weight alone on the deck of the ship, like the guns of some of the Chinese junks of the present day.

Other guns of Henry VIII.'s time.

Other guns belonging to an earlier period than the reign of Henry VIII. have also been recovered from this wreck. They are of iron bars, similar, as regards the principal part of the gun, to that just described, but having a moveable breech. The latter resembles that of the paterara, being slightly conical on the outside, by which means it fitted closely into the bore of the piece, and being introduced when loaded, it was secured in its place by wedges and a block of wood in rear.

French artillery in the 14th century.

Although guns made of bars of iron, with this moveable kind of breech, continued partially in use up to the year when the "Mary Rose" foundered, those in Germany and France were of a superior construction, and as far back as A.D. 1385, bombards were cast to throw a stone shot of 33 lbs. These, owing to their great weight, were transported with extreme difficulty, each on a strong carriage, and being at length conveyed to the required spot, they were placed on the ground by means of a strong crab, near the walls of the place to be attacked. The only kind of carriage then in use, appears to have been simply a block of wood to elevate the muzzle, and another driven into the ground behind the piece to prevent the recoil. The gunners were protected from the enemy's darts by a strong wooden screen on hinges, which was let down for the purpose by one of the men, who, for the sake of safety, was placed in a hole sunk for this purpose on one side of the gun to give him cover. As only four or five discharges could be managed during the day,* other pieces of smaller calibre were fired at intervals to keep the enemy in check. Guns thus supported the machines for launching darts and stones, which were still in use, and on which success so much depended at this period, when artillery was more efficient in the defence, than in the attack of a town or castle. Siege artillery appears to have been much the same in eastern countries also, up to the time of the siege of Constantinople in 1453, excepting that the piece was in this instance, of the enormous calibre of 1,200 lbs., and required in consequence fifty pairs of oxen to drag it, while each discharge occupied two hours.† But owing to the expense as well as the cumbrous nature of the

Mode of serving the guns.

Colossal bombard.

* Etudes sur le passé et l'avenir de l'Artillerie, par le Prince Louis Napoleon Bonaparte, tome ii., p. 62.

† Von Hammer, Histoire de l'Empire Ottoman, tome ii., p. 395.

First formation of a train of siege artillery.

guns, no siege train existed in Europe till 1477, when Louis XI. instead of trusting to one fortress to supply the guns as well as the ordinary machines for attacking another, caused twelve large pieces of ordnance to be cast; viz., three at Paris, three at Orleans, three at Tours, and three at Amiens, partially introducing about the same time the use of metal shot.* But, as the bombards were too cumbrous and the other pieces too light to be serviceable, it was only in the time of his successor, Charles VIII., that artillery became really formidable, and, as means of attack, superseded the use of the catapulta, the trebuchet, the mangonnel, &c. Barrels filled with earth, and subsequently gabions, now protected the besieging batteries. Stones were thrown from mortars instead of bombards, and brass guns cast in one piece with trunnions, replaced the latter: metal shot rapidly fired, took the place of slow discharges of stone shot. In sieges, therefore, owing to rapidity of firing and the increased impetus of shot, the French artillery accomplished as much in the course of a few hours as that of Italy did in as many days.† The short reign of this monarch was on the whole the most important epoch in the annals of modern artillery, which in France was greatly in advance

Brass guns cast in one piece, with trunnions.

* Etudes sur le passé et l'avenir de l'Artillerie, par le Prince Louis Napoleon Bonaparte, tome ii., pp. 97, 111.

† Guicciardini, lib. i., c. xviii.

of the unwieldy cannon of the neighbouring kingdoms of England, Germany, and Italy. Notwithstanding the greatly-increased strength of fortresses, by the addition of earthen parapets mounted with guns, and the introduction of fausse-braies and towers, or even bastions, to give a flanking defence, the concentrated fire of breaching batteries protected by gabions, maintained the superiority of artillery as an offensive weapon in sieges; but, as regarded field operations, little had been done beyond constructing a kind of portable battery, *i. e.*, a four-wheeled waggon, presenting towards an enemy a wooden parapet armed with two large and two smaller guns: it was only after some lapse of time that these were replaced by others of a more efficient kind*

Superiority of artillery as an offensive weapon.

Waggon batteries.

The Musée de l'Artillerie at Paris contains, however, a Turkish culverine which seems to have been one of the earliest light guns that were made in Europe. The tube of this piece is formed of several sections, which are firmly joined to each other by nuts and screws, and it loads at the breech by means of a moveable chamber.

Portable metal cannon constructed.

Besides this, there are two or three other guns, also in sections, and breech-loading, but differing in the mode of accomplishing this latter object.

* Etudes sur le passé et l'avenir de l'Artillerie, par le Prince Louis Napoleon Bonaparte, tome i., p. 69.

Breech-loading field guns.

In one case an iron block or wedge, moved by a notched lever, is used; in another a strong iron screw occupies the interior of the bore; and in a third a sort of plug, put into the latter, is kept in its place by means of a strong screw and horizontal pressure.

Amongst the earliest pieces adapted for rapid movements, were probably those similar to a gun which is in the Royal Military Repository at Woolwich, having been brought from Paris in 1815, and a mortar still shown at Venice with one of its marble shot, said to have been used at the siege of Chioggia in 1385. Both are nearly of similar construction, having a copper cylinder with cords rolled round it, and over this a coating of leather, so as to give a certain degree of solidity, without much exceeding the ordinary size of a cannon in the one case, or that of a mortar in the other. The singular description of guns called *cannons of boiled leather* appears to have continued in use for some time, and those which mainly contributed to the victory gained near Leipzig by Gustavus Adolphus, in 1631, have been thus described:—" They consisted of a thin cylinder of beaten copper screwed into a brass breech, whose chamber was strengthened by four bands of iron. The tube itself was covered with layers of mastic, over which cords were rolled firmly round its whole length; these were equalized by

Light guns made of copper, rope, and leather.

a layer of plaster; and a coating of leather, boiled and varnished, completed the piece. The carriage and the piece were so light, that two men were sufficient to draw and serve this kind of gun, which, as may be imagined, could only bear a small charge."*

Another kind of portable great gun had been known and used in Flanders since 1347, chiefly for fortresses. From the description given by Citadella, in 1387, it would appear that four breech-loading tubes of very small calibre were placed on a two-wheeled cart, with their muzzles protruding through a wooden screen protected by a chevaux-de-frise, which was formed of halberts and pikes, as shown by one of the models in the Rotunda of the Royal Military Repository at Woolwich.

Tube guns placed on carts in

The Scots had something similar, with however only two tubes on each carriage, and the prelates and barons were commanded by Acts of Parliament, 1456 and 1471, to provide such carts of war against their old *enemie* the English.†

A smaller kind of ordnance, called hand-cannon, were introduced into England during the latter year by the Flemings who accompanied

Flanders, and also in Scotland.

* Histoire et Tactique des trois Armes, par Ild. Favé, Capitaine d'Artillerie, Dumaine, Paris, 1845, pp. 80, 81; compared with Meyer, Histoire de la Technologie des Armes à Feu, tome i., p. 77; and Grimoard, Histoire des Conquêtes de Gustave Adolph.

† Henry's Hist. Britain.

Edward IV. on his return from Flanders. They were so small and light that one of them was carried by two men, and fired with a rest on the ground.* It was only, however, during the Burgundian wars that artillery, and that either of the most inefficient or the most cumbrous description, formed part of the equipment of an army. Before this, powerful crossbows capable of throwing stones or iron darts were used, and, as at the battle of Mons-en-Puelle in 1304, were placed in front of the infantry. Occasionally other machines were also employed to throw burning stones as a means of defence.†

The Burgundians use artillery,

The crossbow, &c., was first replaced by the rude fire-arms already noticed, which, being too heavy to be carried by hand, were placed on a cart similar to that used in Flanders about 1342; and from these, small leaden balls or iron darts were discharged.‡

and hand-guns instead of the crossbow, &c.

In 1364 the town of Perouse made 500 hand-cannon, only a palm in length, which were in fact miniature mortars placed in a kind of wooden stock; and in 1382 the French used portable bombards, which were subsequently called culverines. The first field-pieces were placed all round the army, as was the case with the Gantois,

The earlier description of field artillery was small,

* Histoire de France, par le Père Daniel, tome i., liv. vi., p. 321.

† Etudes sur le passé et l'avenir de l'Artillerie, par le Prince Louis Napoleon Bonaparte, tome i., p. 11.

‡ Ibid., pp. 37, 38.

who had 200 cannon carts to defend their position against the Count of Flanders; but being by this arrangement dispersed, and consequently nowhere efficient, the artillery was afterwards placed chiefly in front, and subsequently on the flanks of the line of battle.*

At the battle of Tongres in 1408 both sides had a great quantity of culverines and ribaudequins, which, however, had little or no influence on the result of the engagement;† and three years later the Duke of Burgundy's army, consisting of 40,000 men, had 2,000 ribaudequins, or tubes placed in a row like an organ, and 4,000 cannons or culverines. Artillery did not, however, play an important part in the battles of this period, whether in France or the neighbouring countries. During the Hussite war in Germany the contending armies were encumbered with chariots, which served as a moveable rampart; the cannon of which appear, as at Brux in 1426, at Maleschow in 1424, at Aussig in 1426, and at Tachau in 1431, to have had little or no effect.‡

and produced but little effect in battle.

Under Louis XI. and Charles the Bold, that is in the latter part of the fifteenth century, more efficient organ guns and large cannon were taken into the field; some of the latter

* Etudes sur le passé et l'avenir de l'Artillerie, par le Prince Louis Napoleon Bonaparte, tome i., pp. 44, 45.

† Ibid., pp. 44, 47.

‡ Ibid., p. 49.

threw stones of 120 lbs. weight.* But if a certain degree of importance now belonged to great guns, the portable fire-arms of the Burgundians, the Germans, and the Swiss were, as regards the open field, greatly inferior to the bow and crossbow. Nor is this surprising when it is borne in mind that an expert archer could discharge ten or twelve arrows in the course of a minute, which took effect at an average distance of 200 paces; while the rude hand guns of that age required some three or four minutes for each discharge.†

Arrows were more formidable than imperfect fire-arms.

But a change was commenced about this period by Charles VIII. of France, which was destined to make this arm really serviceable, whether in sieges or battles. In the quaint language of that day, we are told in the Comptes de l'Année, 1489, that "L'Artillerie du roy, marchait toute chargée," or, as it would have been expressed at a later period, "elle marchait chargée sur ses affûts;"‡ that is to say, the important step had been gained of constructing carriages of sufficient strength to bear the recoil of the pieces when fired, and also to serve for their transport from

Gun carriages for service as well as transport,

* Etudes sur le passé et l'avenir de l'Artillerie, par le Prince Louis Napoleon Bonaparte, tome i., pp. 49, 52.

† Ibid., p. 17, compared with Grose's Military Antiquities.

‡ Le Roy en partant du Point-de-l'Arche, fait charger sur l'Artillerie pour faire assaillir Sainte Catherine, en 1449; Mémoires de Jacques du Clery, liv. i., c. xviii., p. 12. as given by Prince Louis Napoleon Bonaparte, vol. i., p. 98.

place to place, with however as yet only the use of two wheels.*

When Charles VIII. determined upon this, and a still more important change, he restricted his artillery to six calibres, giving at the same time an ample proportion of horses to each gun, namely, the double cannon was drawn by 35 horses, the serpentine by 23, the grand culverine by 17, the medium culverine by 7, the large falcon by 2, and the smaller falcon by 1 horse. When marching against Rennes in 1491, Charles brought such a mass of artillery into the field, that 3,000 horses were employed in drawing it.† The whole had field carriages, the double cannon having four wheels, the other pieces only two, and some of the smaller pieces went as fast on a plain as the cavalry.‡ Without going so far (as some have done) as to consider these pieces the commencement of horse artillery, it is clear that guns thus equipped no longer clogged the movements of an army in the field to the same extent as before, so that when invading Italy only three years later, at the head of 30,000 men, he carried into the field, according to the lowest computation, 140 heavy guns, 200 bombards, and 1,000

and horses were introduced by Charles VIII.

Proportion of artillery with the army invading Italy.

* Etudes sur le passé et l'avenir de l'Artillerie, par le Prince Louis Napoleon Bonaparte, vol. i., pp. 95, 96, 99; vol. ii., pp. 113, 114.

† Ibid., vol. ii., pp. 114, 115.

‡ Ibid., p. 100, from Domenichi Venetia, 1608.

Use of hand-guns.

hacquebuttes, or hand-guns, weighing about 50 lbs. each.*

The proportion of artillery differed but little in the next reign. In 1499 there were twenty pieces of artillery for 5,100 men, or four guns to 1,000 men. The organization of this arm being continued by Charles VIII.'s successor, Louis XII. in 1500 was enabled to move his artillery from Pisa to Rome, a distance of about 240 miles, in five days, and his light pieces were sufficiently manageable to be taken rapidly from one point to another during a battle.† When he recovered Genoa in 1507, he had sixty guns of large calibre for an army of about 20,000 men, and by means of his artillery the Venetians were overcome on the Adda in 1509.‡ Francis I. gave a more central organization to the French artillery, having caused 100 large brass guns to be cast at Paris, and 14 magazines to be formed, which contained all the requisites for this arm. He also adopted a shorter and lighter construction for field guns, and the best description of horses were purchased for this service. These guns being quickly drawn to any point, and rapidly served with iron or brass shot, formed a striking contrast to the Italian bombards and other pieces drawn by oxen, and

Francis I. improves the artillery organization, and

* Etudes sur le passé et l'avenir de l'Artillerie, par le Prince Louis Napoleon Bonaparte, tome i., p. 106.

† Ibid., pp. 132, 133. ‡ Ibid., tome i., pp. 132–134; tome ii., p. 117.

formed, as some of them still were, of several sections screwed together, and still firing stone shot.*

Even at this early period the field-pieces of the French were occasionally separated from their heavy guns, although this change was not yet completed: the proportion of artillery, however, to the rest of the army was very considerable. Francis I., when invading Italy at the head of about 33,000 men in 1515, had, besides some of the so-called organ cannon, 74 pieces of ordnance, which, being so placed as to cover the centre of his army by a cross fire, enabled him to defeat a superior force of the best troops in Europe, viz., the Swiss, at Marignan. On this occasion the French artillery played a new and distinguished part, not only by protecting the centre of the army from the charges of the Swiss phalanxes in the way mentioned, and causing them excessive loss, but also by rapidly taking such positions from time to time during the battle, as enabled the guns to play upon the flanks of the attacking columns. Moreover they were ready when required to accompany the cavalry, in order to destroy, by their fire, those obstacles behind which the enemy endeavoured to shelter his men.†

a large proportion of guns brought into the field.

The French field artillery becomes efficient.

The power now attained by the speed and efficiency of well-organized artillery was destined to

Improvement of small arms.

* Etudes sur le passé et l'avenir de l'Artillerie, par le Prince Louis Napoleon Bonaparte, tome ii., p. 115.

† Ibid., pp. 172–180.

First use of light small arms

be contested by a new arm, or rather by a modification of one already known. Two kinds of portable fire-arms had existed for some time, viz., the hand-gun, at first called a culverine, (then a hacquebutte, and afterwards an arquebuse,) and the hacquebutte used with the rest, both of which the Spaniards made sufficiently light to be manageable by hand. Those who used the latter were called hacquebuttiers, and the former arquebusiers, up to the beginning of the sixteenth century, when a change took place. The Spaniards endeavoured to succour Pavia in the face of the French army, in 1524. Francis I. disposed his army in a single line. His numerous artillery, which was placed in front of his troops, commenced the action by a terrible fire simultaneously on the Spanish cavalry and infantry. Both gave way, and the king, perceiving the disorder caused by his cannon in the enemy's centre, advanced at the head of the cavalry, supported by his infantry. The fire of the French artillery being thus masked, the Spaniards rallied and attacked the French, who were thrown into disorder by the fire of 2,000 arquebusiers and 800 musketeers, who now appeared for the first time discharging bullets of two ounces. Thus, what had hitherto been a brilliant victory was snatched from the French monarch, and from this time the use of fire-arms for infantry became an object in Europe.

at the battle of Pavia, and

defeat of Francis I. in consequence.

In the battle of Cerisolles, twenty years later,

Light artillery used in the battle of Cerisolles.

arquebusiers, now called musketeers, shared largely; but more depended on artillery, which on this occasion was generally employed both by the Spaniards and by the French. The former had 18,000 foot, 1,400 horse, and 16 pieces of cannon, and the latter 11,000 infantry, 2,000 cavalry, and 20 pieces of artillery. Three of these were two-pounders, which being drawn by eight horses, accompanied the cavalry, and greatly contributed to the success of the French in this well-contested battle. Each army was formed in a single line, with three intervals, in which the artillery and cavalry were placed, thus showing that there had been much improvement in tactics, particularly as regarded the former arm.* In the war of the Protestant league, the advantages of artillery were not lost sight of, since in 1546, for an army of about 106,000 infantry and cavalry, there were 140 heavy guns, 8,000 arquebusiers, and 300 boats for pontoons. When advancing, the light pieces were in line, occupying the intervals between the squadrons of cavalry, whose manœuvres they always accompanied. In retreating, this arrangement was reversed, the infantry, with the heavy guns, being in front, whilst the cavalry, supported by the field artillery, formed the rear guard.†

* Etudes sur le passé et l'avenir de l'Artillerie, par le Prince Louis Napoleon Bonaparte, tome ii., pp. 189–195.

† Ibid., pp. 197, 198.

The French artillery, however, maintained its superiority, and the use of the prolonge greatly contributed to its efficiency in action. There was also greater simplicity with regard to the kind of guns, which for the field artillery are detailed in a paper of that period, including the order of march.*

Separation of the light from the heavy artillery.

In 1556, when the Emperor Ferdinand marched against the Turks, his heavy and field artillery were separate, the former consisting of 42 heavy, and the latter of 127 light pieces.† Elsewhere some of the latter had attained the greatest efficiency, for at the battle of Remi in 1554, Charles V. employed light guns with limbers, drawn by two horses. These pieces, called the Emperor's pistols, manœuvred at a gallop, and accompanied the movements of the cavalry.‡

Increased importance of artillery in the 17th century.

Artillery, on which so much depended during the latter part of the sixteenth century, became of still greater importance in the early part of the seventeenth, when three distinguished leaders, Henry IV. of France, Maurice of Nassau, and Gustavus Adolphus of Sweden, gave their attention to this arm as an important branch of the art of war.

* Etudes sur le passé et l'avenir de l'Artillerie, par le Prince Louis Napoleon Bonaparte, tome ii., pp. 185–189.

† Ibid., pp. 210, 211.

‡ Commentaires des dernières Guerres en la Gaule Belgique, par François de Rabutin, liv. ix., p. 481.

Small arms as well as artillery improved by Henry IV. of France.

The first followed in this particular the system of Henry II., using however improved missiles, such as tin cases filled with steel bolts or darts, also canvas cartridges filled with small balls, and hollow shot filled with combustible materials. But the introduction of small arms also occupied his particular attention, especially for the cavalry. Pistols had been given to some of the latter, and a short arquebuse to another portion, since known as Carabineers.* And as regarded the protection of the infantry from the enemy's cavalry, a complicated square formation was adopted, by which some 2,500 men, chiefly pikemen, were placed in a body, composed of four small squares, having arquebusiers in the intervals, and again other arquebusiers placed in small squares at the exterior angles, and also opposite to the centre (outside) of the great square; the fire of the latter however at this period, about 1611, was only efficient at a very short distance.†

Modifications in artillery adopted by Maurice of Nassau.

The second of these leaders, Maurice of Nassau, adopted for his artillery the system of the Spaniards, with however only three calibres instead of four, but, as in their case, amply horsed. Except that two of the pieces were too heavy, being 48 and 24 pounders, the equipments differed little from those of the present day, having limber-boxes

* Etudes sur le passé et l'avenir de l'Artillerie, par le Prince Louis Napoleon Bonaparte, tome i., pp. 250, 259, 265, 267.

† Ibid., p. 255.

and other improvements in the carriages; and his gunners attained great facility in working the pieces.*

Gustavus Adolphus organizes his army.

But the genius of the third individual, Gustavus Adolphus, extended to every branch of his army, of which an uniform mode of clothing was the least important part. His infantry was divided into regiments and companies, and the latter again into sections and squads; his army was formed into divisions, each comprising portions of infantry, cavalry, and artillery, and every division being under a particular commander, was, whether on the march or in action, complete of itself, and ready for service. In addition to this, an improved fire-arm, now known as the musket, of a lighter construction and carrying a smaller ball, was also adopted.

He improves his artillery, and

Not satisfied with the improvements which had taken place in the German artillery about the beginning of the seventeenth century, he introduced important changes by the creation of calibres of 3, 4, 6, 12, and 30 pounders, all cast of a lighter construction than heretofore; and by the adoption of the use of cartridges with shot attached, these pieces might be discharged eight times before the musket could be fired six.

adopts a new order of battle.

The order of battle adopted by Gustavus was, for the most part, in two lines, with intervals in

* Etudes sur le passé et l'avenir de l'Artillerie, par le Prince Louis Napoleon Bonaparte, tome i., pp. 270, 271, 272, 277.

the first line to admit of the advance of the second in case of necessity; his artillery was in the centre of the first line, protected by cavalry and platoons of infantry, and ready to move. The second line was composed of the rest of his troops, with the baggage in rear.* At the battle of Leipzig, in 1631, his victory over Tilly was complete. Gustavus Adolphus had about 30,000 men and 72 guns, against 35,000 and an equal proportion of artillery under Tilly. The former arranged his forces in two lines, the one in advance having the infantry in the centre, with 30 pieces of artillery in front. The infantry was divided into very small sections, for the purpose of executing the most rapid movements without disorder; the cavalry, also in small bodies, was on the wings, with some musketeers in reserve. The second line consisted of three brigades of infantry, and seven squadrons of cavalry, with two of the latter in reserve. The Imperialists, whose infantry and cavalry were disposed in large unwieldy masses, occupied the chain of hills near Leipzig, with their artillery in the rear.

Battle of Leipzig, A.D. 1631,

The action commenced with a cannonade, which lasted two hours, and which, owing to the superior calibre of the guns of the Imperialists, was favourable to Tilly; but his descent at the

commenced by artillery

* Geschichte des dreissigjährigen Kriegs von Schiller, 9 band. Stuttgart and Tubingen, 1838, p. 75.

Success of Gustavus Adolphus.

head of his centre, to take advantage of this circumstance, was opposed and repulsed by the fire of the 30 pieces of artillery in the centre of the Swedish line. Tilly then united with his right and fell on the Saxons with such impetuosity, that they took flight on the first shock. The king, nothing daunted, assembled a mass of artillery to resist Pappenheim's attack on his right wing, and by the rapid fire of his light leather guns, completely defeated him, notwithstanding his vigorous and often-repeated attacks. He then, at the head of the greater part of his army, took possession of the heights previously occupied by the Imperialists, and captured and turned their guns against themselves, thus obtaining a brilliant victory which opened all Germany to the conqueror.*

Daring passage of the Lech by Gustavus Adolphus.

Tilly, however, endeavoured to defend the entrance to Bavaria, at the passage of the Lech, behind which he was strongly posted; but the power of artillery triumphed. Four batteries of 72 pieces were so placed by Gustavus at a bend, and on a higher bank of the river, as to produce a cross-fire on the opposite bank, and a bridge was thrown over the swollen river under this powerful protection: this operation was ineffec-

* Geschichte des dreissigjährigen Kriegs von Schiller, pp. 224-228, compared with Histoire et Tactique des trois Armes, par Ild. Favé, Capitaine d'Artillerie, pp. 81-84, and Etudes sur le passé et l'avenir de l'Artillerie, du Prince Louis Napoleon Bonaparte, tome i., pp. 326-331.

tually opposed by Tilly with an inferior number of guns, and the Imperialists were eventually put to the route.* The death of Tilly, in this engagement, placed the distinguished Wallenstein at the head of the Imperialists, and after a short campaign of manœuvres, the contending forces met at Lutzen. Wallenstein was strongly posted with the cavalry on his flanks, and a double line of tirailleurs, supported by artillery, in his front. Gustavus formed his forces as usual in two lines, arranged nearly as at Leipzig, having 66 cannon, 26 of which, of heavy calibre, protected his centre. A cannonade of two hours was succeeded by a charge of the Swedes, which was followed by a series of attacks and manœuvres during a protracted struggle. The battle continued doubtful long after the death of Gustavus Adolphus, and the Imperial artillery fell three times into the power of the Swedes. It was recovered the second time by the opportune arrival of Pappenheim, with a body of cavalry; but the final advance of the second line of the Swedes, with their reserve artillery, again put them in possession of the enemy's guns, and left them masters of the field. The artillery of the Imperialists remained on the ground. This memorable battle was distinguished by the moveableness of the Swedish troops, and

Battle of Lutzen and death of Gustavus Adolphus.

* Geschichte des dreissigjährigen Kriegs von Schiller, pp. 275, 276, compared with Etudes sur le passé et l'aveuir de l'Artillerie, par le Prince Louis Napoleon Bonaparte, pp. 331, 332.

more particularly by that of Gustavus' artillery, which was at all times ready to be brought up rapidly, so as to concentrate an overwhelming fire wherever defensive protection might be required, or an impression was to be made upon the enemy's position; whereas, the less manageable cannon of the Imperialists being comparatively immoveable, were of little service during movements in advance, and became even injurious when the troops were repulsed during the latter operations.*

Defective state of the imperial artillery.

From the result of the battles thus cursorily noticed, it appears that, although retaining too many calibres, the artillery of Gustavus Adolphus was admirably organized, embracing as it did limbers, carrying cannister shot and other kinds of ammunition ready for action, and, what was no less important, having the allotment of a proportion of reserve artillery, in addition to that destined to accompany the troops during their movements in action. Moreover it will be seen that this distinguished commander was the first who fully appreciated the importance of causing the artillery to act in concentrated masses, and who well understood the saving of life consequent on taking into the field a due proportion of this arm.

Advanced state of the artillery of Gustavus Adolphus.

* Schiller, pp. 344-353, compared as above; Tactique, &c., p. 86; Etudes sur le passé, &c., tome i., pp. 336-341.

CHAPTER IV.

PRESENT STATE OF CONTINENTAL AND BRITISH ARTILLERY.

Light Artillery introduced by Frederic the Great.—Construction of the Prussian, Austrian, and Russian Field Artillery.—Strength of the respective Armies.—Organization of the French Artillery Service.—Strength of the French Army, and proportion of Artillery in the Field.—British Artillery in the Peninsula.—Troop of Horse Artillery and Field Battery.—Proportion of Artillery to an Army.—Organization of Field Artillery in India.—Strength of the Indian Army.—Proportion of Officers to the Men in the Artillery of various European Services.—Importance of a due proportion of Artillery, as shown in several battles.—Marengo, Austerlitz, Jena, and Wagram.—Also in Napoleon's Campaign in Russia, at Lutzen, Leipzig, and Battle of Waterloo.—Want of Artillery complained of by the Duke of Wellington.—Military strength depends upon an adequate force of Artillery.—Deficiency of Artillery in India.—Holkar's Guns destroy the British Artillery at Mahidpore.—Battles of Chillianwallah and Goojerat.—Human Life economised by the use of Artillery.—Number of Guns in the British Colonies.—Imperfect Defence of the Coast of Great Britain.—Troops and Artillery greatly required in case of an Invasion.—Number that would be required for Defence.—Exposure to Attacks by Steamers.—Difficulty in protecting British Commerce. —Possibility of Invasion.—The Duke of Wellington's opinion of the danger to which England is exposed from a Steam Flotilla.—Possibility of the latter escaping British Cruizers.—Speed and safety in Landing Troops on the English Coast.—Additional means of Defence proposed.

Frederic the Great introduces light artillery.

PERCEIVING the great disadvantages caused by unwieldy artillery to the opponents of Gustavus Adolphus, Frederic the Great adopted such pieces as were suited to accompany and even to form an integral part of the rapid manœuvres he had introduced, as part of that system of tactics which required a greater proportion of artillery than heretofore, and which so largely influenced throughout Europe the organization and proportion of this service relatively to the other arms.

It is manifest that guns drawn by manual labour, or imperfectly horsed, must retard the operations in the field, and become at times worse than useless in battle; and hence the self-evident necessity of field artillery capable of rapid movements.

In adverting to the equipments of this portion of an army, precedence may fairly be given to that kingdom which owed its consolidation to the royal warrior just mentioned.

The Prussian artillery of the present day.

A Prussian battery of artillery consists of eight pieces, viz., six guns and two howitzers, with, when complete, the following details:—

—	A Troop of Horse Artillery					A Field Battery.				
	Guns.	Howitzers.	Officers.	Men.	Horses.	Guns.	Howitzers.	Officers.	Men.	Horses.
Guns	6	..	..	..	..	6	..	..	..	..
Howitzers	..	2	..	..	..	..	2	..	..	..
Officers	..	..	4	..	..	..	..	4	..	..
Surgeon	..	..	1	..	..	..	..	1	..	..
Non-commissioned Officers	..	..	..	11	..	..	..	..	10	..
Bombadiers	..	..	..	16	..	..	..	..	16	..
Trumpeters	..	..	..	2	..	..	..	..	..	..
Drummers	..	..	..	..	..	..	..	..	2	..
Gunners	..	..	..	114	..	..	..	..	108	..
Artificers	..	..	..	..	..	..	..	..	2	..
Drivers	..	..	..	19	..	..	..	..	19	..
Horses	..	..	..	..	222	..	..	..	..	114
Total	6	2	5	162	222	6	2	5	157	114

One troop of horse artillery and four companies of field artillery, or 40 guns, form a division of artillery under a field officer, and three divisions, or 120 guns, form a brigade under a brigadier, which is the complement for a corps d'armée of about 30,000 men; and of these grand subdivisions of the Prussian army there are, including that of the guards, nine, or 291,448, being one gun to about 270 men. In 1833 the Prussian army appears to have consisted of 248,000 infantry, 43,448 cavalry, and 39,150 artillery, &c., or 330,598 in all.* But in 1850 and 1851, the army, including the Landwehr, exceeded 400,000 men.

Proportion of guns to the Prussian army.

Austria, the ancient as well as the modern rival of Prussia, had prepared a reorganization of her artillery, combining all the advantages derived from the experience of the gigantic wars which convulsed the Continent during the early part of the present century. This change, however, was not carried out until the conclusion of the Hungarian war.

Organization of the Austrian artillery.

The result has been an improved field and rocket artillery, without any horse artillery, the latter being supplied by what is called a cavalry or flying battery; in which, instead of being mounted, the gun detachments are carried on a

Cavalry or flying battery.

* Prussia in 1833, translated from the French of M. de Chambray. Boone, London, 1834.

Field artillery and its

light spring waggon provided with cushions, called a würst; and as in Prussia the batteries consist of eight pieces of ordnance, viz., six guns and two howitzers.

—	Field Battery.						Cavalry Battery.						Rocket Battery.				
	Guns.	Howitzers.	Officers.	Men.	Carriages.	Horses.	Guns.	Howitzers.	Officers.	Men.	Carriages.	Horses.	Tubes.	Officers.	Men.	Carriage.	Horses.
Guns	6	.	.	.	.	.	6	.	.	.	.	.	.	.	.	.	.
Howitzers	.	2	.	.	.	.	.	2	.	.	.	.	.	.	.	.	.
Tubes	.	.	.	.	.	.	.	.	.	.	.	.	12	.	.	.	.
Captains	.	.	1	.	.	.	.	.	1	.	.	.	.	1	.	.	.
Lieutenants	.	.	3	.	.	.	.	.	3	.	.	.	.	3	.	.	.
Non-commissioned Officers	.	.	.	15	.	.	.	.	.	15	.	.	.	.	16	.	.
Trumpeters	.	.	.	2	.	.	.	.	.	2	.	.	.	.	2	.	.
Waggon Masters	.	.	.	16	.	.	.	.	.	16	.	.	.	.	16	.	.
Gunners and Drivers	.	.	.	131	.	.	.	.	.	141	.	.	.	.	170	.	.
Artificers	.	.	.	5	.	.	.	.	.	5	.	.	.	.	6	.	.
Waggons	.	.	.	.	8	.	.	.	.	.	24	.	.	.	.	.	.
Horses	.	.	.	.	.	150	.	.	.	.	.	180	.	.	.	.	86
Total	6	2	4	169	8	150	6	2	4	179	24	130	12	4	210	.	86

Proportion to the rest of the army.

A regiment of artillery consists of 24 batteries, or 192 guns, with six companies in reserve without guns; and there are five regiments, or 960 guns, in addition to 15 rocket batteries, having 180 rocket tubes, with two companies besides in reserve. As the aggregate force in time of peace is about 378,552 men, there is nearly one gun, or one rocket tube, to 332 men.

The Russian artillery.

Formerly the Russian field batteries had 12 guns, whether they were heavy or light. The former consisted of eight 12-pounder guns with

Existing organization.

four 20-pounder howitzers, and the latter of eight 6-pounder guns with four 10-pounder howitzers. But the troop of horse artillery had only eight guns, viz., four 6-pounder guns and four 10-pounder howitzers. In 1832 and 1834, however, the batteries were reduced to eight pieces, the light being of the same number and calibre as those of the horse artillery: the heavy batteries having four 12-pounder guns and four 20-pounder howitzers. Besides the preceding, some of the Cossack regiments use 3-pounder howitzers, which, however, are not found to produce much effect against an enemy.

The Russian field artillery.

The following is the equipment of the Russian field service:—

	Troop of Horse Artillery.						Field Battery.					
	Guns.	Howitzers.	Officers.	Men.	Carriages.	Horses.	Guns.	Howitzers.	Officers.	Men.	Carriages.	Horses.
Guns	6	2	..	..	..	..	6	2	..	..	..	..
Captains, 1st and 2nd . . .	..	..	2	..	..	..	..	..	2	..	..	..
Lieutenants and Sub-Lieutenants	..	..	4	..	..	..	..	..	4	..	..	..
Fireworkers .	..	..	..	8	..	..	..	..	..	8	..	..
Surgeons . .	..	..	1	..	..	..	..	..	1	..	..	..
Sergeants . .	..	..	..	8	..	..	..	..	..	8	..	..
Bombadiers . .	..	..	..	16	..	..	..	..	..	16	..	..
Gunners . . .	..	..	..	76	..	..	..	..	..	74	..	..
Drivers . . .	..	..	..	48	..	..	..	..	..	46	..	..
Trumpeters . .	..	..	..	2	..	..	..	..	..	2	..	..
Ammunition Waggons . .	..	..	..	..	8	..	..	..	..	..	8	..
Baggage Waggons	..	..	..	..	3	..	..	..	..	..	3	..
Horses . . .	..	..	..	..	..	209	..	..	..	..	..	142
	6	2	7	158	11	209	6	2	7	154	11	142

Strength of the Russian army and artillery, &c.

One heavy and two light batteries, under a superior officer, form a brigade of field artillery, and two troops of horse artillery, also under a superior officer, a brigade of horse artillery. To each corps d'armée is attached one division of artillery, composed of two brigades, each consisting of four batteries of eight pieces of ordnance, one brigade of horse artillery, and one reserve battery, which, together with the park of artillery, and three battalions of sappers, amount to 6,000 men, and 120 pieces of ordnance for a corps d'armée of 60,000 men; or 360,000 men and 720 pieces of ordnance for the whole army, which amounts in time of peace to 594,000, being one per cent. of the whole population of Russia.

Organization of the French artillery.

The field artillery of the French is of three kinds, viz., batteries à cheval, batteries montées, and foot batteries, which nearly correspond with those in our service, each troop or battery having, as we had in war-time, six pieces of artillery, two however being howitzers instead of only one, as in our service.

—	Batterie à Cheval.					Batterie Montée.					Batterie à Pieds.				
	Guns.	Howitzers.	Officers.	Men.	Horses.	Guns.	Howitzers.	Officers.	Men.	Horses.	Guns.	Howitzers.	Officers.	Men.	Horses.
Captains, 1st class	.	.	1	.	3	.	.	1	.	3	.	.	1	.	2
Ditto, 2nd class	.	.	1	.	3	.	.	1	.	3	.	.	1	.	2
Lieutenant, 1st class	.	.	1	.	2	.	.	1	.	2	.	.	1	.	1
Ditto, 2nd class (Sous-lieutenant)	.	.	1	.	2	.	.	1	.	2	.	.	1	.	1
Sous-officier, as Adjutant	.	.	.	1	1	.	.	.	1	1	.	.	.	.	.
Maréchal Logis Chef, or Sergeant-major	.	.	.	1	1	.	.	.	1	1	.	.	.	1	.
Sergeants	.	.	.	8	8	.	.	.	8	8	.	.	.	8	.
Fourier, Pay-sergeant	.	.	.	2	2	.	.	.	2	2	.	.	.	1	.
Brigadiers, or Corporals	.	.	.	12	12	.	.	.	12	6	.	.	.	12	.
Artificers	.	.	.	6	6	.	.	.	6	.	.	.	.	6	.
Cannoniers' Servants, 1st class Gunners	.	.	.	28	66	.	.	.	24	180	.	.	.	66	.
Cannoniers' Servants, 2nd class Gunners	.	.	.	54		.	.	.	36		.	.	.	100	.
Cannoniers' Conducteurs or Drivers, 1st class	.	.	.	40	156	.	.	.	44		.	.	.	.	.
Ditto, ditto, 2nd class	.	.	.	54		.	.	.	66		.	.	.	.	.
Artificers in wood and iron	.	.	.	4	.	.	.	.	4	.	.	.	.	4	.
Farriers	.	.	.	3	3	.	.	.	3	3	.	.	.	.	.
Collar-makers	.	.	.	2	.	.	.	.	2	.	.	.	.	.	.
Trumpeters	.	.	.	3	3	.	.	.	3	3	.	.	.	2	.
Guns	4	.	.	.	.	4	.	.	.	.	4	.	.	.	.
Howitzers	.	2	.	.	.	.	2	.	.	.	.	2	.	.	.
Grand Total	4	2	4	218	268	4	2	4	212	214	4	2	4	200	6
Enfants de Troupe	.	.	.	2	.	.	.	.	2	.	.	.	.	2	.

Working proportions of the troops and batteries.

A troop, a battery, and a company, are each commanded by a captain of the first class, and divided into three sections, one of which is in charge of the capitaine-en-second, and the other two under the two lieutenants. The horse battery, as the name implies, has mounted detachments. In the second or mounted battery, the gunners are carried, like ours, on the limbers and ammunition waggons, and in the third case, they walk, and even drag the guns by manual labour.

Sixteen companies or batteries form a regiment, of which there are 14, besides one of pontonneers, in addition to a park train of four squadrons of eight companies each for the field, besides the garrison and coast artillery.

The superior officers of a regiment are—

Superior officers of a regiment of artillery.

1 colonel.
1 lieut.-colonel.
7 chefs d'escadron, who are usually detached.
1 major of the rank of chef d'escadron, to take charge of the accounts.
1 captain-instructor in equitation and driving.
2 adjutants-major, with the rank of captains.
1 officer, superintendent of clothing, ditto.
1 treasurer, ditto ditto.
1 assistant to ditto, with rank of lieutenant or second lieutenant.
1 principal surgeon.
2 assistant ditto.
1 veterinary ditto.
2 assistant veterinary ditto.

	Batteries à Cheval.	Batteries Montées.	Batteries à pied.		Total.
For each of the four first regiments there are	3	10	3	=	16
For each of the other ten regiments	2	10	4	=	16

Making, for the 14 regiments, 32 batteries à cheval, 140 batteries montées, and 52 foot batteries, or 224 batteries in all, with, as already shown, four officers to each. The 15th, or pontoon regiment, has the same staff as the other fourteen, with the exception of having four instead of eight chefs d'escadron, and of being without a captain-instructor or a veterinary officer. But each of the twelve companies of which it is composed has, like the batteries, four officers.

Number of regiments and batteries, &c.

Besides the several officers included in the preceding organization, the artillery service has an extensive staff of its own, viz. :—

General staff of the artillery service.

6 generals of division } making part of the general staff
13 generals of brigade } of the army.
31 colonels.
31 lieut.-colonels.
35 chefs d'escadrons.
112 captains of the 1st class, and
75 ditto, separately employed in permanent situations in the arsenals and garrisons.

The proportion of artillery relatively to the two other arms varies between two and three guns for 1,000 men, according to the composition of the army, and the nature of the country in which it is to operate.

In 1822 the military force of France consisted of eight corps d'armée of about 40,000 men, each

Strength of the French army,

comprising two divisions, and each of the latter composed of two, or sometimes of three brigades. To 10,000 men are allotted—

	8-Pounders.	Howitzers.
1 Battery of Horse Artillery	4	2
3 ,, Mounted	12	6
1 ,, of Horse Artillery in reserve .	4	2
1 ,, Mounted	4	2

and proportion of artillery in the field.

And for the corps d'armée the following pieces:—

—	12-Pounders.	8-Pounders.	Howitzers.	Total.
For the 16 Divisions { 48 Batteries à Pied .	64	192	96	288
For the 16 Divisions { 16 ,, in Reserve		..	32	96
16 Batteries à Cheval	..	64	32	96
16 ,, in Reserve	..	64	32	96
Total . 96 Batteries . . .	64	320	192	576

And as a general reserve—

—	12-Pounders.	8-Pounders.	Howitzers.	Total.
6 Batteries of Divisions	8	24	12	36
2 ,, Reserve		..	4	12
2 ,, Divisions	..	8	4	12
2 ,, Reserve	..	8	4	12
Total . 12 Batteries . . .	8	40	24	72

The field equipment is supposed to have two-thirds guns and one-third howitzers, about one-sixth of the whole being of large calibre. The guns are 12-pounders, 8-pounders, and 4-pounders, and the howitzers are respectively of 22 centimetres (8⅝ inches), of 16 centimetres (6¼ inches), of 15 centimetres (5·9 inches), and of 12 centimetres (4·72 inches).

ORGANIZATION of a MOUNTAIN BATTERY—MATERIAL.

	In the Field.	In the Reserve Park.	
12-pounder howitzers	6	..	
Carriages	7	1	8, giving 2 spare.
Portable forge	1	..	
Ammunition boxes—			
For cartridges for howitzers	42	82	8 rounds per box, of which 1 is of ball. In all 16 rounds per howitzer, 20 of which are of ball.
For cartridges for infantry	10	20	1,000 cartridges per box; 30,000 in all.
For tools and spare articles	4	8	
Containing the portable forge	2	..	
Pack saddles	42	56	
Mules	44	56	100 in all, on account of 1 for each gun, 1 for each carriage, 1 for two boxes, and 2 spare.

Owing to its effects on distant bodies of men, and to its facility in overcoming those obstacles which give security against musketry, mountain artillery has acquired great importance during the campaigns in Algeria. Without such means, the Arabs might have successfully defended their towns, villages, and positions against the French, and caused immense loss to the invaders. To avoid difficulties of this kind it became absolutely necessary that every column should be accompanied by guns to keep the Arabs at a distance; and this system was followed up till the arrival of General Bugeaud, who adopted the arrangement of keeping the guns in reserve, and only bringing them forward in case of a reverse Mountain artillery.

or to produce a decided effect during an engagement.

Objections made in France to the employment of artillery officers on the staff, and decision in their favour.

In addition to the extensive staff belonging to this branch of the service, so large a share of general employment falls to the artillery officers in France, as to cause the envy of their cotemporaries of the line. With reference to the undue favour supposed to be bestowed on this service since the time of Napoleon, General Prival published a voluminous pamphlet, in which he broadly states that officers of the line are alone fit to command a corps d'armée, from which he would exclude those of the artillery, who should, he says, be confined exclusively to their own service. In consequence of the view so strongly taken by certain generals of infantry and cavalry, a report prevailed that it was intended to confine officers of the artillery and engineer services to their own specific duties; but the following reply appears to have set the question at rest:—

"C'est, en effet, une étrange prétention que celle qui consiste à soutenir qu'un officier d'artillerie, qui a du nécessairement faire des études plus approfondies que les autres, dont on exige beaucoup plus de connaissances, &c., est par cela même inférieur aux autres; rien assurément ne justifie une pareille assertion; et je passe au fait qu'indépendamment de ses connaissances en général beaucoup plus variées, un officier d'artillerie est infiniment mieux au courant des services des autres armes, que les officiers d'infanterie et de cavalerie ne le sont de ce qui concerne l'artillerie."

The French artillery, although less in proportion than that of Prussia and Russia, is still very considerable, and an addition to this branch of the service appears to be in contemplation at this moment.

The increase of 1840 raised the artillery service to 30,604, the cavalry being 58,294, and the infantry 257,454, or, in all, 346,152 men.* Strength of the three arms in 1840.

The vicissitudes of a very trying service necessarily caused the effective strength of the British field artillery to vary; but the full equipment in the Peninsula, of the troop then commanded by Lieutenant-Colonel, now Lieut.-General, J. Webber Smith, was 6 guns, 17 carriages, 5 troop officers, 1 assistant-surgeon, 16 non-commissioned officers, 90 gunners, 2 trumpeters, 7 artificers, and, including 30 additional men from the driver corps, 90 drivers, or, in all, 211 officers and men. A company of artillery, then attached to a car brigade of 9-pounders, under the command of Lieutenant-Colonel, now Lieutenant-General Sir Robert Gardiner, K.C.B., consisted of 6 guns, 13 carriages, 5 company officers, 1 driver, and 1 medical officer, 20 non-commissioned officers, 120 gunners, 2 drummers or trumpeters, 7 artificers, and 80 drivers, in all 236 officers and men. Strength of a troop of British horse artillery, and of a field battery in the Peninsula.

* The artillery details have been obtained from the Archives du Depôt Central d'Artillerie; and the number of cavalry and infantry is taken from the Mémorial d'Officiers d'Infanterie et de Cavalerie d'après les Documents Officiels, &c., pp. 1, 2, 3. Strasbourg, 1846.

Some modification took place after the return of the army of occupation from France; and the following details will show the equipment which was then recommended for the field in time of war:—

Details of a troop of horse artillery and field battery as subsequently proposed.

	Six-Pounder Troop of Horse Artillery.*						Nine-Pounder Battery.†					
	Guns.	Howitzers.	Spare and other Carriages.	Officers.	Men.	Horses.	Guns.	Howitzers.	Spare and other Carriages.	Officers.	Men.	Horses.
Guns	5	..	..	..	..	..	5	..	..	..	..	..
Howitzers	..	1	..	..	..	..	..	1	..	..	..	..
Spare Forge and other Carriages	..	..	15	..	..	..	..	..	15	..	..	..
Captains	..	..	..	1	..	..	..	..	..	1	..	..
Second Captains	..	..	..	1	..	..	..	..	..	1	..	..
First Lieutenants	..	..	..	3	..	..	..	..	..	2	..	..
Second Lieutenants	..	..	..	..	..	..	..	..	..	1	..	..
Surgeon or Assistant Surgeon	..	..	..	1	..	..	..	..	..	1	..	..
Staff Sergeants	..	..	..	..	2	..	..	..	..	..	1	..
Sergeants	..	..	..	..	4	..	..	..	..	..	4	..
Corporals	..	..	..	..	6	..	..	..	..	..	6	..
Bombadiers	..	..	..	..	6	..	..	..	..	..	6	..
Gunners and Drivers, including Servants and Batmen	..	..	..	..	127	..	..	..	..	..	130	..
Trumpeters and Drummers	..	..	..	..	1	..	..	..	..	..	2	..
Farriers	..	..	..	..	1	..	..	..	..	..	1	..
Carriage Smiths	..	..	..	..	1	..	..	..	..	..	1	..
Shoeing Smiths	..	..	..	..	3	..	..	..	..	..	2	..
Collarmakers	..	..	..	..	2	..	..	..	..	..	2	..
Wheelers	..	..	..	..	2	..	..	..	..	..	2	..
Horses	..	..	..	..	..	156	..	..	..	..	..	116
Total	5	1	15	6	155	156	5	1	15	6	157	116

* Remarks on the Organization of the Corps of Artillery in the British Service; London, printed for Rowland Hunter, 72 St. Paul's Churchyard, 1818, p. 185. Also, Report of the Sub-committee of Officers of the Royal Artillery in February, 1820.

† Remarks on the Organization, &c., p. 187; and Report of the Commissioners for Inquiry into Naval and Military Promotion, &c., 1840, p. 82.

Proportion of artillery for an army of 27,000 men.

It is understood that the sub-committee of officers who were nominated in 1819 and 1820 to make a revision of the artillery equipments, namely, Colonels William Millar, Sir William Robe, K.C.B., Sir George Wood, G.C.H., and Lieutenant-Colonels Sir Augustus Fraser, K.C.B., and Sir Alexander Dickson, K.C.B., proposed that to each brigade of about 4,000 men one battery of six guns should be attached, and the same number to a division of cavalry of about 3,000 men. This, including two troops of horse artillery, and one field battery in reserve, would give for a corps d'armée of 24,000 infantry and 3,000 cavalry, 60 guns, or 10 batteries; three of these to be horse artillery, and having besides a proportion of small arms, ammunition, and artillery waggons, with also an 18-pounder battery, giving the following details:—

—	Cavalry and Infantry.	No. of Batteries.	No. of Guns.	No. of Howitzers.	Reserve Ammunition Waggons for the Artillery.	Ammunition Waggons for Small Arms.	Artillery Horses.
Cavalry Division	3,000	1	5	1	3	4	218
Two Brigades of Infantry, No. 1 Division	8,000	2	10	2	6	12	398
,, ,, No. 2 Division	8,000	2	10	2	6	12	398
,, ,, No. 3 Division	8,000	2	10	2	6	12	398
Reserve Artillery Field Batteries . .	..	3	15	3	40	60	914
,, 18-pounder Battery, Officers, N.C.O., and Gunners of the Artillery Field Batteries	2,288	1	3	1	3	..	200
	29,288	11	53	11	64	100	2,526

—	Troop of Horse Artillery.						Field Battery.						Golundauz Battery.				
	Guns and Howitzers.	Carriages, including Spare.	Officers.	N. C. O. and Men, Europeans.	Gun Lascars, &c.	Horses.	Guns and Howitzers.	Carriages, including Spare.	Officers.	N. C. O. and Men, Europeans.	Gun Lascars, &c.	Horses.	Guns and Howitzers.	Carriages, including Spare.	Officers.	Natives.	Bullocks.
	Troop of horse artillery.						Ordinary field battery.						Golundauz battery.				
Guns and Howitzers .	6	..	..	..	..	..	6	..	..	..	..	..	6	..	..	..	..
Gun and Spare Carriages	..	15	..	..	..	..	..	15	..	..	..	..	..	15	..	..	..
Captains	..	..	1	..	..	..	..	..	1	..	..	..	..	..	1	..	..
First Lieutenants . .	..	..	2	..	..	..	..	..	2	..	..	..	..	..	2	..	..
Second ditto	..	..	1	..	..	..	..	..	1	..	..	..	..	..	1	..	..
Surgeon or Assistant-Surgeon	..	..	1	..	..	..	..	..	1	..	..	..	..	..	..	..	..
Staff Sergeants . . .	..	..	..	1	..	..	..	..	..	1	..	..	..	..	..	..	..
Sergeants	..	..	..	7	..	..	..	..	..	7	..	..	..	..	..	..	..
Corporals	..	..	..	6	..	..	..	..	..	6	..	..	..	..	..	..	..
Bombadiers . . .	..	..	..	10	..	..	..	..	..	6	..	..	..	..	..	..	..
Trumpeters or Drummers	..	..	..	4	..	..	..	..	..	2	..	..	..	..	..	2	..
Gunners	..	..	..	60	..	..	..	..	..	60	..	..	..	..	..	..	..
Farriers	..	..	..	2	..	..	..	..	..	1	..	..	..	..	..	..	..
Rough Riders . . .	..	..	..	2	..	..	..	..	..	..	..	..	..	..	..	..	..
Saddlers	..	..	..	1	..	..	..	..	..	..	..	..	..	..	..	..	..
Subadars or Jemadars .	..	..	..	..	..	..	..	..	..	..	1	..	..	..	..	3	..
Havildars	..	..	..	..	1	..	..	..	..	..	2	..	..	..	..	9	..
Naiks . , . . .	..	..	..	..	2	..	..	..	..	..	2	..	..	..	..	8	..
Privates (Native Lascars)	..	..	..	..	24	..	..	..	..	..	62	..	..	..	..	100	..
Total . .	6	15	5	93	27	..	6	15	5	83	67	..	6	15	4	122	..

The preceding table (p. 88) shows the organization of the field artillery in India, the basis of which was furnished to the East India Company by the late Sir Alexander Dickson, G.C.B. The Bengal service is given. The other two Presidencies only differ in one or two immaterial details.

Organization of the field artillery in India.

SUPERIOR OFFICERS.

Proportion of superior officers of artillery.

For a Brigade of Five Troops, Bengal Establishment.*	For a Battalion of Artillery, Bengal Establishment, Five Companies.	Battalion of Golundauz of Eight Companies.
1 Colonel.	1 Colonel.	1 Colonel.
1 Lieut.-Colonel.	1 Lieut.-Colonel.	2 Lieut.-Colonels.
1 Major.	1 Major.	2 Majors.
5 Captains.	5 Captains.	8 Captains.
10 First Lieutenants.	10 First Lieutenants.	16 First Lieutenants.
5 Second ditto.	5 Second ditto.	8 Second ditto.
1 Sergeant-Major.	1 Sergeant-Major.	1 Adjutant.
1 Quartermaster-Sergeant.	1 Quartermaster-Sergeant.	1 Surgeon.
		2 Assistant ditto.
		1 Sergeant-Major.
		1 Quartermaster-Sergeant.
		1 Subadar-Major.
		14 Subadars.
		30 Jemadars.
		120 Havildars.
		120 Naiks.

Men.

Total of the Indian army and artillery.

Total of the artillery service	15,719 and 7,309 followers.
Cavalry	20,000
Infantry	180,000
Total of the army . . .	200,000 and 59,784 followers.

The following shows the result of the preceding tables as regards the proportion of officers and men of the continental field artillery; also that of the British artillery in Europe as well as in India, exclusive of field and superior officers.

* There are but four companies of gunners; the fifth company has officers without men.

Tabular view of the proportion of officers to the men, in the several continental as well as in the British services.

	Officers.		Non-com. Officers and Men.		Officers.		Non-com. Officers and Men.		Officers.		Non-com. Officers and Men.		Officers.		Non-com. Officers and Men.
The Prussian Troop of Horse Artillery has .	4	to	162	or	1	to	40	And the Field Battery has	4	to	157	or	1	to	39
The Austrian Troop of Cavalry Artillery has	4	to	179	or	1	to	44	The Mounted Battery has	4	to	169	or	1	to	42
The Russian Troop of Horse Artillery has .	6	to	158	or	1	to	26	The Mounted Battery has	6	to	154	or	1	to	25
The French Troop of Horse Artillery has .	4	to	218	or	1	to	54	The Batterie Montée has	4	to	212	or	1	to	53
Troops of British Horse Artillery in Spain had	5	to	205	or	1	to	41	Battery, then called Car Brigades, had . .	6	to	229	or	1	to	38
Troop of Horse Artillery, as organized in 1820, had	5	to	155	or	1	to	31	Battery in 1840 had . .	5	to	157	or	1	to	31
Troop of Horse Artilllery in India, present establishment, has . . .	4	to	120	or	1	to	30	Field Battery in India at present has . . .	4	to	150	or	1	to	37
Troop of Horse Artillery, as proposed in this work	3	to	108	or	1	to	36	Field Battery as proposed in this work . . .	3	to	100	or	1	to	33

This tabular view of the continental and British services, will at once show that the project of the writer gives rather more officers in proportion to the men, than there were serving in the Peninsula, and also considerably more than there are in any continental artillery service at present, the Russian excepted.

Use of light field batteries.

In briefly reverting to the progress of artillery, it will be recollected that scarcely a century has elapsed since 3, 6, 12, and 24 pounder guns were ordered to be cast of a sufficiently light construction to advance in battle with the troops to which they belonged;* and the result of the seven years' war was greatly to increase the proportion of artillery to the other arms; at the same time it received a fresh organization, by which it was enabled to act in masses, instead of delivering the weak and desultory fire of battalion guns.

Number of field guns fixed by Napoleon.

"Artillery in the present day," observed Napoleon at St. Helena, "decides the fate of nations, and he who knows how to bring a mass of this arm to act rapidly upon a particular point, unknown to the enemy, is sure to carry it." The number of field guns fixed by the Emperor in 1811 for France, was 1,224 horsed, and the same number in reserve, exclusive of garrison guns. But, observes Paixhans,† this number, which in France

* History of my Own Times, by the King of Prussia, vol. i., p. 75.

† Constitution Militaire de la France, &c., pp. 307, 308.

is limited on account of the expense, is greatly exceeded in Prussia, Austria, and Russia, where the cost, particularly of horses, is much less.

The system of tactics that prevailed before the time of Gustavus Adolphus, or rather that which became better known under Frederic the Great, when an army moved, as it were, in one piece, involved the necessity of encumbering it with supplies to such an extent as to restrict, if not altogether to cripple its operations. But a subdivision into divisions and corps d'armée, each of which was complete as to organization, and to a certain extent independent, enabled the army safely to occupy more space than it had hitherto done: this, under ordinary circumstances, facilitated the means of obtaining supplies, and the general was thus enabled to avail himself of the ordinary routes without spreading the order of march. The use of two ranks for the cavalry and three for the infantry necessarily extended the line of battle, and facility of movement, consequent upon an improved organization, permitted without danger the use of intervals between the brigades and divisions of the army. Thus battles occupied an extensive space; each corps, excepting as regarded the security of its flanks by the neighbouring divisions, engaging in a separate combat. Under these circumstances the various batteries played an important part in the several divisions to which they belonged. But

Corps d'armée, &c. organized.

Troops occupy more space.

this was only what might be considered the little wars of artillery; for although used with powerful effect in many battles, this arm was only employed in detached sections, to the direction of which the functions of the general of artillery were confined, instead of the scientific direction of large masses, as was done subsequently by the great genius who gave to this arm the efficiency and velocity on which his success so largely depended.

Artillery attached to divisions of an army.

In adopting a system, as it were, of separate centralization, Napoleon gave to his corps d'armée the means of acting by itself. Each had its reserve and park of artillery; and, not having any fixed place in the order of battle, could move independently to accomplish any particular object, and afterwards take any special position that might be desirable to enable it to share in the battle. Such an arrangement gave a manifest advantage over the extended order of other continental armies, since Bonaparte could, even with inferior forces, vigorously assume the offensive; and by breaking through some part of the enemy's line with a superior force, attack him in rear and flank, and also continue this attack against another division, without giving time for the adversary to be aware of the check he had received, or to perceive the danger of his retreat being cut off, by having more hostile troops poured

Napoleon's organization

for offensive warfare.

Napoleon's plan of attacking the enemy.

through the interval thus made. A battle of this kind was not general along the whole line, but became a series of engagements at particular points, often distant from one another; and the course followed by Napoleon to obtain a key to the enemy's position, previous to a general attack, appears to have been different to that which has been imagined from the apparent eagerness with which he fought.

Before forming his plan of the intended battle, he allowed that portion of his army which was nearest to the enemy almost to commence a general engagement, in order that the exact position and purpose of the adversary might be ascertained. The reinforcements which were frequently importunately demanded by his generals at this juncture were often withheld, and the temporary success of the enemy was disregarded. But when there was no longer any doubt as to the position and force of the enemy, Napoleon's plan of attack, which was then speedily formed and energetically executed, generally depended upon a great effort to carry one or two points under a preponderating fire of artillery. But instead of employing his whole force, as has been supposed, the attack was commenced by a portion of his troops only, which he continued to strengthen until the enemy's reserve had become engaged. The decisive moment having then arrived, Napoleon's reserve was brought up fresh, and having passed through the opening in the enemy's line, one portion attacked him in flank and in rear, whilst another endeavoured to cut off his retreat. It was at this juncture that the whole French army assumed the offensive, and the victory was frequently gained, when perhaps three-fourths of the line were ignorant how it had been accomplished.*

The artillery accompanied his movements.

As the basis of decisive attacks, it was absolutely necessary that the artillery service should unite efficiency to such rapidity of movement as would enable this arm to share in the most difficult and complicated manœuvres of battle; and when thus organized, it can rarely fail to secure the success of an attack made on any particular point, or to recover lost ground, if this arm is brought up in a mass to act at a critical moment, such as during the first part of the celebrated action of Marengo. The battle commenced by opening a fire from 100 pieces of cannon,† which proved irresistible; and the French army, being broken and defeated, had already retreated before 27,000 Imperialists under General Melas, when the engagement, then actually lost by the French, was renewed under altered circumstances.

Marengo at first gained by artillery, and afterwards lost by its absence.

Bonaparte, perceiving the approach of Dessaix,

* Histoire et Tactiques des trois Armes, et plus particulièrement de l'Artillerie de Campagne, par Ild. Favé, Capitaine d'Artillerie, pp. 198, 402, 403.

† Précis des Evènements Militaires, compared with Alison's History of Europe, vol. iv., p. 336.

Napoleon becomes the assailant.

opportunely brought up fifteen guns under Marmont, which having opened a destructive fire, checked the advance of an Austrian column, and at this moment taking advantage of a defile, he first rallied his flying troops, and then became the assailant. This happened at the very moment when some of the Austrian cavalry had been detached on another service, while the infantry being then at some distance from the main body of artillery, were actually preparing to avail themselves of the opening supposed to have been made through the enemy to continue their march to join General Wucassartch on the Adige. Thus the laurels which General Melas had apparently won, were snatched from his brows by the presence of mind and daring intrepidity of his skilful opponent.*

Battles of Austerlitz, Friedland, &c.

The services of the artillery were scarcely less important during the great battles of Austerlitz, Jena, and Friedland. In the latter, thirty-six pieces of cannon opportunely brought up by General Lénarmont, did more than had been accomplished by 20,000 men under Ney, assisted by the division of Dupont.† In these contests, and others which may be noticed, the artillery having

* Histoire et Tactique des trois Armes, &c., par Ild. Favé, Capitaine d'Artillerie, pp. 214–216, compared with Histoire des Guerres des Gaulois et des Français, par Joseph Servan, Général de Division, tome v., p. 273, &c.

† Histoire et Tactique des trois Armes, &c., par Ild. Favé, Capitaine d'Artillerie, pp. 231–237.

moved and fought in masses, produced the most decided effects.

Napoleon's preparations.

Wagram is at once a striking example of the importance of artillery, and, at the same time, one of the most remarkable battles in the history of the world. The bridges across the Danube having been destroyed, that part of the French army which remained in the island of Lobau, was separated from the main body; Napoleon had therefore no other resource than either to sacrifice part of his army, or to undertake one of the greatest and most difficult operations ever accomplished in war, viz., forcing a passage in the face of a powerful army, commanded by one whose talents and experience almost equalled his own.

Passage of the Danube.

Everything turned upon speed, and so great were the exertions made, that during the early part of the night of the 5th of July, 1809, one complete bridge was swung into its place by the current, and five others being completed simultaneously, Bertrand was enabled to say to his master, "Sire, il n'y a plus de Danube." The passage commenced forthwith, under the protection of 109 pieces of cannon on the right, in addition to those on the island of Lobau, and, in the morning, the plain near the left bank was occupied by the French army with 400 guns. The Austrians, on their part, had entrenched the

The Austrian position.

villages of Enzersdorf, Essling, and Gros-asperm, forming a strong position, from which they advanced and gained advantages, not only on their right and centre, but on the left also. Attaching most importance to the latter on account of the proximity of the Archduke John's army, the Emperor, in person, endeavoured to drive the enemy back from this point, but a grand battery secured this part of the line, and victory appeared to be gained by the Archduke Charles.

This trying juncture required all the skill of the Emperor, and a series of bold manœuvres followed, the result of which was a double attack made simultaneously on the centre and left of the Austrians, with nearly the whole French force. Davoust opened a fire on the latter point, with 50 pieces of artillery, and having made some impression, partly outflanking, and separating the enemy from the Archduke John,

Double attack by Napoleon.

Napoleon seized this moment to make, under cover of the fire of 110 guns, a grand attack on the centre of the line, which, after the most desperate efforts on each side, was at length successful, and the Archduke Charles, finding his centre forced, ordered a retreat.*

Napoleon enters Russia with 1,372 guns.

Following the example of the Austrians, as well as that of his predecessor, Frederic the Great, in increasing the proportion of his artil-

* Histoire et Tactique des trois Armes, par Ild. Favé, pp. 243-248.

lery, Napoleon entered Russia with 1,372 guns; and at the great action of Borodino, which was mainly a battle of artillery, and the greatest fought since its invention, the fire of upwards of 1,100 pieces of cannon caused, on both sides, a frightful carnage, which was continued until the Russians abandoned the field of battle. On this occasion Napoleon's army numbered 133,000 men, with 590 guns, and that of the Emperor of Russia 132,000,* with 640 guns, being one piece for 225 in the former, and for 206 men in the latter case.

1,230 guns in the battle of Borodino.

Rising above the destruction of his forces in Russia, Napoleon prepared for another campaign, with a fresh army of conscripts; when, as he approached Lutzen under his altered circumstances, he propounded the new maxim in war: "Qu'une bonne infanterie soutenue par de l'artillerie doit savoir se suffire."†

Importance of artillery.

This he exemplified by the formation of columns of infantry covered by artillery; for at the moment when the battle was all but lost, an attack made by the Imperial Guard, preceded by a concentrated fire from a battery of eighty pieces of artillery under General Drouot, broke through the allied Russians and Prussians under their respective sovereigns; the retiring troops were

Order of battle at Lutzen.

* Précis des Evènements Militaires, compared with Alison's History of Europe, vol. viii. p. 696.

† Bulletins Officiels de la Grande Armée, tome ii., p. 160. Paris, 1824.

then vigorously pursued notwithstanding the French inferiority of cavalry.

The passage of the Elbe, and

Napoleon, on finding the passage of the Elbe disputed by the retreating forces, called out in a voice of thunder to General Drouot, "One hundred pieces of cannon!" and eighty guns disposed along the heights of Preisnitz, speedily secured a passage, which was then effected by means of a bridge of rafts,* towards the strongly entrenched position of Bautzen behind the Spree. Masses of artillery being disposed on every projection commanding the opposite bank of this river, bridges were speedily established, and the next day a succession of obstinate contests at different points terminated by Napoleon's grand attack on the heights of Kreekwitz, which, being aided by 100 pieces of cannon, was crowned with victory;† but like the preceding battle it was incomplete, owing to the want of cavalry.

artillery at Bautzen.

Battle of Leipzig.

In a subsequent and still more fearful struggle, when the fate of Europe was contested near Leipzig, 200 pieces of artillery preceded the advance of the allied armies on the 16th of October, and later in the day 1,000 were brought into action. The contest continued until the centre of the assailants was broken by the concentrated fire of 150 guns under General Drouot. At the moment

* Bulletins Officiels de la Grande Armée, tome ii., pp. 172, 173.
† Ibid., pp. 200–202.

of the crisis on the 18th, 800 pieces of allied artillery crowned a vast semicircle of ridges, of which Wachau was the centre: from this commanding position a concentric fire was discharged on the French army, and the 500 pieces of artillery by which it was covered.

Relative strength in guns and men of the contending armies.

The allies numbered 280,000 men, with about 1,370 guns, in this gigantic battle; and the French had 166,000 men, with 720 guns, being one gun in proportion to 203 men for each army. The French in this prodigious struggle for empire are said to have discharged 230,000 rounds.*

Battle of Hanau.

General Wrède intercepts the French.

Were it necessary, numerous instances of the same kind might be mentioned in addition to the preceding during the late wars; but as they relate to successful operations, it may be desirable to notice one of the reverse kind, namely, that in which General Wrède endeavoured to intercept the French army at Hanau. Napoleon, on being told that it was impossible to force a passage, sent Drouot to examine the state of things, who speedily reported that with fifty guns, supported by two battalions of the Old Guard, he would open a passage. Napoleon hastened to the spot, and asked how the guns would be placed to effect this object. Drouot replied, that he would place twelve guns, without caissons, on the road, and three others elsewhere; and having by these means

* Bulletins Officiels de la Grande Armée, tome ii., p. 335.

attracted the enemy's attention, the remaining guns would form successively on the right of those already formed. Fifty pieces thus partly placed on the flank of the enemy's artillery produced a decided effect. The Bavarian cavalry now attempted to charge the guns, but being repulsed during this operation by the cavalry of the Guard, the guns took a fresh position, so as to cannonade the left of the Austro-Bavarian army, which was driven back on Hanau, and the retreat of the French secured in consequence.

The French force a passage.

Proportion of guns and men in the continental armies, and

In 1833 Prussia had 27,000 artillerymen, with 864 guns, or 1,080, including the landwehr;* and the proportion of guns taken into the field by the allied Russian and Prussian armies, varied from between 1 for 159 men, at Lutzen, to 1 for 203 men at Leipzic, the latter being also the proportion of the French in the same battles.

the British army at Waterloo.

When the British army assumed the offensive on the afternoon of the 18th June, 100 pieces of cannon crowned the heights of Waterloo, and covered the advance.† But in this memorable conflict, when the proportion of artillery was so much greater than it had previously been in our service, there were but 156 British guns to contend against 246 French, or, for those who were actually engaged, only 1 gun for about 408

* Prussia in 1833, translated from the French of M. de Chambray, p.64.

† Précis des Evènements Militaires, compared with Alison's History of Europe, vol. x. p. 955.

men,* whilst, during the Peninsular war, the proportion was considerably less; thus, in the action at Vimeira (18 guns and 16,000 men), there was 1 gun to 889 men; at Corunna (12 guns and 14,000 men), 1 gun to 1,166 men; at Talavera (100 guns and 59,997 men), 1 gun to 600 men; at Albuera (38 guns and 30,000 men), 1 gun to 789 men; at Salamanca (24 guns and 20,000 men), 1 gun to 833 men; at Vittoria (90 guns and 80,000 men), 1 gun to 889 men; at Orthès (48 guns and 37,000 men), 1 gun to 771 men; and at Thoulouse (64 guns and 52,000 men), 1 gun to 812 men; or, taking a general average, there was 1 gun to 785 men.†

Proportion of the guns in Spain.

Writing from Villa Toro, October 18, 1812, the Duke of Wellington reminds Lord Bathurst "that in the battle of Salamanca, the French army had more than twice the number of his pieces of ordnance;" his Grace adds, that instead of eight brigades of 9-pounders, there should be nearly double that number, or about 90 pieces.

Marked deficiency of the proportion of British artillery in the field.

In a subsequent Despatch, January 27, 1813, the Duke observes, that "the equipment of ordnance is infinitely lower than that of any army now acting in Europe."

As military strength in the field greatly depends

* Siborne's History of the War in France and Belgium, vol. i., pp. 377, 461.

† Taken from Napier's History of the Peninsular War.

The efficiency of an army mainly depends on a due proportion of artillery.

upon the efficiency of this arm, it may be safely considered as an axiom in war, that the smallest army which can be efficient in the field is that to which a due proportion of artillery is attached, of at the same time sufficient calibre.

The deficiency of guns throughout the protracted war in Spain has just been shown, and our army had the additional disadvantage of having generally only 6-pounders to oppose to the French 8-pounders, which nearly answer to our 9-pounders.

Deficiency of British artillery in India.

Up to the recent great battles, it has been the custom in India also to take the field under similar disadvantages. At Mahidpore, for instance, in 1817, the enemy had 32 pieces of cannon of various calibres, between 8-pounders and 18-pounders, with 31 smaller, between a 7-pounder and 2-pounder, in all 63, to which 20 small guns, chiefly horse artillery, and 12 foot artillery, were opposed. The army under General Sir Thomas Hislop advanced to attack Holkar, who was strongly posted on the opposite banks of the Sippra river, with numerous guns commanding the principal ford. A troop of horse artillery, under a distinguished officer, Captain Noble, was pushed forward to engage the enemy, whilst the cavalry and infantry were preparing to cross. Holkar's heavy guns* were so well served, that the

Holkar's guns at Mahidpore,

* Two of 18, two of 17, four of 16, two of 14, and one a 2-pounder.

eight light pieces of the horse artillery were speedily put hors de combat, and Captain Noble made known to the Commander-in-Chief "that he might bring on the infantry as soon as he pleased, for the guns were knocked to pieces." **greatly outnumbered the British artillery.**

But notwithstanding this unequal combat, in which 63 horses were killed on the spot, Holkar's position was eventually carried, and his guns taken, though after a severe loss. **Result of the battle.**

Chillianwallah is another striking example of the disadvantage of attacking an army covered by a powerful artillery, without an adequate proportion of this arm to support the infantry and cavalry. **Battle of Chillian-wallah,**

Lord Gough had on the field ten 18-pounders, fifteen 9-pounders, thirty 6-pounders, six 12-pounder howitzers, in all seventy-two guns; but only a portion of these had been brought up when the enemy opened a fire from the greater part of his artillery. The latter appears to have been well served and judiciously placed to cover his position. It consisted of sixty pieces, of which twelve were taken, viz., six 7-pounders, one 7½-pounder, two 6-pounders, one 6½-pounder, one 5½-pounder, and one 3-pounder. **and the British and Sikh artillery.**

The costly mistake on this occasion was carefully remedied, and with the most brilliant results, in the succeeding battle near Goojerat. On this **Battle of Goojerat,** occasion Lord Gough brought into the field ten 18-pounders, twenty 9-pounders, forty-five 6-pounders, with eight 8-inch, four 24-, and nine 12-pounder howitzers, or ninety-six pieces in all. To these, fifty-nine pieces were opposed by the Sikhs, of which fifty-three were captured, viz.:—

and the Sikh guns.

One 18-pounder.	Three 7-pounders.
Two 16-pounders.	Six 6-pounders.
One 12-pounder.	One $3\frac{1}{2}$-pounder.
Five 9-pounders.	Two 2-pounders.
Nineteen 8-pounders.	Eight howitzers and three mortars.
Two $7\frac{1}{2}$-pounders.	

In this glorious struggle, which may be considered the Waterloo of India, the advance of Lord Gough's forces was effectually covered by his numerous and well-served artillery. The 18-pounders, each of which was drawn by a couple of elephants, quickly took advantageous positions, and were of the greatest service in silencing the enemy's guns. It was in fact almost entirely a battle of artillery before the infantry was much exposed. Goojerat, therefore, takes a high place in the annals of British warfare, by proving in the most unquestionable manner, *how greatly life is economised by the judicious use of a powerful artillery.*

Heavy guns drawn by elephants.

Life economised by the use of artillery.

This, for such an empire as Great Britain, is very important; and considering our present inferiority in artillery, it is worthy of attentive consideration whether it is not absolutely necessary that there should be an augmentation of this arm both at home and abroad, for garrison and field service. However pressing the former may be, there is, owing to our maritime superiority, less danger in neglecting it than the latter, since the organization of an artillery force requires much time and instruction. Our numerous colonies demand a serious augmentation of artillerymen. Gibraltar, for instance, has 653 guns mounted, for which the five companies stationed in that fortress could not furnish quite one man for each gun. At Malta there are 486 guns mounted, with three companies of artillery, or two men to three guns. In the Ionian Islands, 351 guns, with three companies, less than one man to each gun. In America, the West Indies, and other colonies, there are 1,928 garrison guns,* with (after deducting those in the batteries) twenty-eight companies to man them, or not quite two men to each gun.

Number of guns in the British colonies,

and great deficiency of gunners.

But if our colonies are inadequately provided with the personnel of artillery, the growing power of steam has made a more effective armament of our coasts an object of vital consideration. For

Imperfect defence of

* Report on the Numerical Deficiency, Want of Instruction, and Inefficient Equipment of the Artillery of the British Army, with suggestions for its partial re-organization, respectfully addressed to Lord Seymour and the Committee of the House of Commons on the Army, Navy, and Ordnance Estimates, by Major-General Sir Robert Gardiner, K.C.B., Royal Artillery. Published in Jones's Woolwich Journal for July, 1848.

9,100 miles of sea-coast.

the protection of some 9,100 miles of sea-board in Great Britain and Ireland* there are only about 1,523 guns, which, few as they are, and supposing only the present number of field batteries to be manned, would have scarcely three gunners to each piece.

That England has been considered vulnerable in this respect, will be evident from the following passage, translated from Paixhans' work :—

Great Britain is exposed to

"Instead of constructing ships of the line to prepare victories for the English, let us, on the contrary, build light fast vessels, such as will give the greatest scope to the powers of steam and artillery. Let these be entrusted to those energetic men who are to be found in our fleet and army, and let them depart from ten different ports, so as to arrive the same night and hour at the same place on the English coast; and having, either by disembarking or by a shower of shells, inflicted a fearful and long-to-be-remembered blow, they should repeat a similar attack, sometimes at 100, sometimes at only ten leagues from the former point; whilst other vessels, in open sea, fall unexpectedly on some of those rich convoys whose value is the life of British commerce."†

It was stated by the Duke of Wellington, in a letter dated the 9th January, 1847, in answer to the observations of the Inspector-General of Fortifications (Lieutenant-General Sir John Burgoyne,‡ K.C.B.) on the possible results of a war with

* Following the sinuosities of the coast, but omitting the islands, there is a distance of 9,100 miles to defend; viz., for England 3,700, Ireland 2,900, and Scotland 2,500.

† Constitution Militaire de la France, pp. 168, 169.

‡ Published in Jones's Woolwich Journal, February, 1848.

France under our present system of military preparation, that—

"The whole force employed at home, in Great Britain and Ireland, would not afford a sufficient number of men for the mere defence and occupation, on the breaking out of war, of the works constructed for the defence of the dockyards and naval arsenals, without leaving a single man disposable."

In another part of the same letter, his Grace observes—

"We are not safe for a week after the declaration of war. . . . I was aware that our magazines and arsenals were very inadequately supplied with ordnance and carriages, arms, stores of all denominations, and ammunition. . . .

an attack by steam vessels.

"You will see, from what I have written, that I have contemplated the danger to which you have referred. I have done so for many years. I have drawn to it the attention of different administrations at different times. . . . .

Unprotected state of England.

"I quite concur in all your views of the danger of our position, and of the magnitude of the stake at issue. I am especially sensible of the certainty of failure if we do not, at an early moment, attend to the measures necessary to be taken for our defence, and of the disgrace—the indelible disgrace—of such failure."

In another paragraph the possibility of danger is thus summarily explained:—

"I know of no mode of resistance, much less of protection, from this danger, excepting by an army in the field capable of meeting and contending with its formidable enemy, aided by all the means of fortification which experience in war and science can suggest."

We likewise learn from the same unquestionable authority that, after providing the requisite garrisons for Portsmouth, Devonport, &c., only 5,000 men of all arms could be put under arms, if required for any service whatever in the field. The invading force is supposed to be only 40,000 men, whose "embarkation with their horses and artillery might take place at the several French ports on the coast, and their disembarkation at named points on the English coast." *

Deficiency of troops and artillery, in case of an invasion.

The lamentable deficiency of the field artillery of Great Britain, with reference to such an emergency will be sufficiently evident, if we bear in mind that there are only 52 guns horsed for service in Great Britain; viz., five troops of horse artillery, and eight batteries at Woolwich and elsewhere. Whereas, if the number were to be based on that of the continental armies, for instance, on the Prussian corps of 40,000 men assembled on the Meuse, in 1815, with 200 pieces of cannon,† there should be 178, or, according to the limited allowance of the Sub-Committee of Artillery,‡ 79 guns for the 35,612 regular troops in Great Britain, without providing any whatever for an additional force; or even failing this, for the militia and volunteers.

Troops for the defence of Great Britain,

Less than three corps, each of 50,000 men, could not be considered an adequate protection with reference to invasion; viz., one in Ireland,

* Letter of his Grace the Duke of Wellington to the Inspector-General of Fortifications, January 7, 1847.

† Letter of the Duke of Wellington to Earl Bathurst. Brussels, April 15, 1815.

‡ Ante, page 86.

and two in Scotland and England, one of the corps in these countries being allotted for the coast defences, and another kept in reserve, to be assembled by railway at some central point in the country. The smallest number with which the protection of Great Britain and Ireland could be undertaken, would, according to the Duke of Wellington, be a force, including militia, of 150,000 men; which, allowing three guns to every 1,000, would require 450 guns, or at the low estimate of the Artillery Committee, 333 guns to be brought into the field. To horse such a number, in order to provide against a possible contingency, is scarcely to be thought of, more particularly as, in case of emergency, large assistance in point of untrained animals would be at command. As in the case of the rest of the army, a numerical force of artillery is in these times greatly increased by the means of rapid locomotion, since a short time would suffice to concentrate it, not only at any one particular place, but even at several points in succession. The available force, however, could not be beyond the actual number of guns and troops that ougtht to be assembled at any one point of attack. It is true that by means of railways the guns could be sent to occupy certain positions, and thus to act, though less efficiently, with a small proportion of hired horses, or even without any at all; but it is evident that

and proportion of guns.

Railways could convey guns to certain points.

Horses might be obtained, but

in this case it would be absolutely necessary to send experienced gunners to serve them. Horses, that would be useful to a certain extent, could be obtained and hastily trained; but this is absolutely out of the question with regard to the gunners. If it be true, as has been stated, that something may be done with inferior cavalry or infantry, but that bad artillery is worse than useless, the possibility of providing a sufficient number of well-trained artillerymen for field service, on such an emergency, becomes an object of paramount necessity. And the force estimated by the illustrious Commander-in-Chief to be requisite for the protection of the country would call for 9,713 artillerymen, or about 3,000 men in addition to what we now have, supposing every gunner to be taken into the field for this purpose, thus leaving the *garrisons and sea-batteries to be manned by the Coast Guard and volunteers.* An increase of 3,000 men to the service seems, therefore, to be indispensable, on the broad ground that though there might possibly be time to call out the militia and raise volunteers for a sudden emergency, *the necessary instruction could not be given to the additional gunners, who are required to support and assist such a force in the defence of the kingdom.*

9,713 gunners would be required in addition to the Coast Guard.

An invasion would fail in the end.

With her maritime position, the finest fleet in the world, and numerous steamers to protect the coast, as well as the means of assembling her land

Danger to Great Britain.

forces by railway in twenty-four hours, England need have but little anxiety about the ultimate result of a sudden attack. But when we meet with the following passage, in addition to that already quoted from Paixhans—" In future England will have to learn that although she will doubtless be able to defend herself, yet the security derived from her hitherto inaccessible position as a country has received a serious shock, and that she may, in her turn, know what it is to tremble for her firesides; and this will be an immense step for France, and for the rest of the world" *—it behoves us to inquire whether the means at present exist of even a moderately good defence, in case of any of the estuaries or great commercial arteries of Great Britain being suddenly attacked? A reply in the negative has already been given, and the brochure of the Prince de Joinville is to the same effect:—

Constant exposure to the attacks of steamers.

"With steamers," he observes, "an aggressive warfare of the most audacious nature may be carried on at sea. We are then certain of our movements—at liberty in our actions: the weather, the wind, the tides, will no longer interfere with us, and we can calculate clearly and with precision. The most unexpected expeditions are possible. In a few hours, armies may be transported from France to Italy, Holland, Prussia. What has been once accomplished at Ancona with rapidity, aided by the wind, may be again done" [as against Rome]

* Constitution Militaire de la France, p. 169.

"without such assistance, and even in spite of it, and with still greater quickness."*

And elsewhere it is stated—

"Our successes would not be transcendant, because we should be careful in compromising our whole resources in any one decisive meeting; but we should wage war with advantage, because we should attack two points equally vulnerable, namely, the confidence of the British people in their insular position, and her maritime commerce. Who can doubt but that, with a well-organized steam-navy, we should possess the means of inflicting losses and unknown sufferings on an enemy's coast, which has never hitherto felt all the miseries that war can inflict?"

Again—

Difficulty in protecting the commerce of Great Britain.

"Our steam-navy would then have two distinct spheres of action. First, the Channel, where our own harbours might shelter a considerable force, which, putting to sea in the obscurity of the night, might attempt most numerous and well-organized attacks. Nothing could hinder this force from reuniting at a given point on the British coast before daylight, and there it might act with impunity."

The Duke of Wellington's opinion of the

Anticipating, as some of our naval commanders have done also, that time and chance would at length permit one of the supposed flotillas to elude our blockading squadrons, and reach some part of the British coast, the Duke of Wellington, before the Committee of the House of Commons

* Naval Strength of France in comparison with that of England, translated from the French of his Royal Highness the Prince de Joinville by W. Peake, Esq., 4to edition. London, Parker and Furnivall, 1844, p. 5.

on Shipwrecks, stated, with reference to that part of our coast immediately opposite to France,—"In the event of war, I should consider that the want of protection and refuge which now exist, would leave the commerce of this part of the coast, and the coast itself, in a very precarious situation."

But the more decided opinion expressed by his Grace in his letter of the 7th January, 1847, to Major-General Sir John Burgoyne, should be conclusive. "This discovery [steam] immediately exposed all parts of the coasts of these islands which a vessel could approach at all, to be approached at all times of tide, and in all seasons, by vessels so propelled, from all quarters. *We* are in fact assailable, and at least, liable to insult, and to have contributions levied upon us on all parts of our coasts."*

mischievous power obtained by an enemy by steam-vessels.

Since it cannot be denied that the loughs and bays in Ireland, the firths in Scotland, and the estuaries as well as the bays in England, are at least very imperfectly, and, generally speaking, not at all protected in case of attack: our present means of defence, being inadequate for *both* objects, would either be employed on the exposed points, or, if concentrated with reference to the more effective defence of the interior, would leave the

Difficulty in defending an extensive coast line.

* The remarkable letter from which this is an extract was first perused by the writer in the 'Friend of China,' in 1848, having appeared in the Hong-Kong papers, as well as in those of all the rest of the world.

former almost wholly unprotected. If the coast be made, as it probably would be, the leading object, the whole force will be required on or in the vicinity of the sea-board. If, on the contrary, the greater attention be given to the means of meeting an enemy in the field, the coast must be denuded of troops, or at best only partially defended, in order that, by leaving one-third of the troops and guns for the protection of the north of England and Scotland, the remainder, or about 24,000 men and 36 guns horsed, and about 50 more without horses, may be assembled at some point in the south of England. This can scarcely be considered a mere speculation, since it must be admitted that a fleet of steamers may eventually find an opportunity of suddenly landing on our shores, in a few hours, a force double that which was recently transported with such speed from Toulon to the coast of Italy. It was no disparagement to the untiring vigilance of our blue-jackets that a formidable army was landed in Egypt in 1798, or that another was only prevented by the elements from effecting the same thing in Ireland; nor will it be any reproach to our navy if the first lull after a storm which clears the channel for a moment, should enable a hostile flotilla to reach our coast, and disembark with a degree of speed and safety hitherto impossible in maritime operations: for to effect the latter object,

Rapidity of landing a force by steamers,

after avoiding the blockading squadron.

it would only be necessary that two small-sized steamers should be run ashore broadside on, which being done, planks on one side, and vessels coming up successively on the other, would form a bridge in a moment for the enemy; not, in fact, the first since the Norman conquest who will have reached the coast of Britain.

Steam-vessels may be considered as floating bridges to any point of the English coast.

Thus viewed, steam is scarcely less than a floating bridge, which may have one extremity at any one of the various ports of the Continent situated between the Baltic and the harbour of Cadiz, and the other on our own shores; where, however, the threatened point may and can be defended (provided we have the means of doing so), whenever a passage across the channel is attempted. With reference to subsidiary means, it might, were this the place, be easily shown that, without the consumption of time and the vast expense required by the construction of regular fortifications, defensive works might be executed at a comparatively small expense; such works, whether for the protection of particular points on the coast, or for that of a great central depôt in the interior, could not be mastered without heavy artillery; and the transport of the latter would give all that England requires—a little time. This equally applies to what might be done for the protection of the capital, and the great arsenal of the empire in its vicinity, which may be con-

Additional fortifications.

sidered branches of the same important object; considering their proximity to the coast, it is not too much to say that means should be taken for their temporary security, were it only for two or three days; and, in connexion with such precautions, we may mention the great assistance that would be afforded for defensive warfare by the hedgerows and enclosures of England when compared with such means in other countries.

Means of defensive warfare.

CHAPTER V.

ON THE STATE OF THE BRITISH ARTILLERY.

Backward state of Promotion during the late War.—Number of the Artillery at Waterloo.—The Officer in Command was the oldest man in that Army.—Sketch of the early state of the Royal Artillery.—Proportion of Artillery with the British Army in Germany in 1760, in Portugal and in Germany in 1762.—The Artillery duty performed by Battalions in 1763.—Artillery sent to India in 1791.—Strength and Equipment of the Artillery in 1793.—Drivers hired to conduct the Guns.—Numerous Artillery brought into the Field by the French in 1793.—State and Number of the British Artillery at the beginning of the Nineteenth Century.—Reduction consequent on the Peace, and hopeless state of this Arm.—Comparison with other Services.—Disadvantages with regard to Pay and Allowances.—Comparison of the Royal Artillery with the Engineers, and also the Madras Engineers.—Slow Promotion caused by a want of due proportion of Company to Field Officers.—Strength of a Company in the Field during the War.—Remedies proposed.—Other projects have hitherto failed.—Rank of the Guards once offered to the Artillery, and declined.—The Company system and its objections.—Defects of the Artillery.—Change proposed by the late Col. Sir Augustus Fraser, K.C.B.—Reorganization of the East India Company's Artillery by Lord Moira.—Proposed change to small Companies and Battalions.—Batteries of Eight Guns proposed by the Duke of Wellington.—Advantages of this arrangement.—The ninety-six Companies to become 144.—Changes that this would cause Abroad and at Home.—Advantages of such alteration.—State of the Regiment in former times as to Reliefs.—Colonial Service fell to one part of the Regiment exclusively, and Active Service to another.—Part of the Companies of Artillery were kept at Woolwich and the rest Abroad.—Distribution of one of the Battalions of Artillery in 1813. Strength of the Companies of Artillery in various years.—Average Strength of the Adjutant's Detachments at Woolwich in War-time.—The latter would have been sufficient to give a partial relief to the Companies stationed abroad.

State of the Royal Regiment of Artillery during the war.

With regard to the present state of the artillery, it need scarcely be observed that, previously to the peace, one of the great defects in the artillery service in this country was, as it still is, from the second captains upwards, a want of promotion among the officers; as a proof of which it may be stated, that two meritorious captains of the artillery, Johnstone and Festing (both now retired), were first lieutenants of five years' standing at Vimeira, and served with the same rank in almost every subsequent battle up to that of Thoulouse; whilst their schoolfellows in the line were leading regiments to glory in the very same actions.

It is well known, also, that the allied artillery, amounting to about 8,000 men, and 6,000 horses and mules,* was, throughout a great part of the Peninsular war, under the command of a first captain (Dickson), whose cotemporaries were then serving as major-generals in command of divisions of the same army.

7,500 men and 5,800 horses employed with the artillery at Waterloo.

The greatest struggle in which England has as yet been engaged, viz., the battle of Waterloo, found this talented and experienced officer, who had served in the warfare of the Peninsula, still a captain serving under Sir George Adam Wood, a brevet colonel, who had at the moment 7,500 men and 5,800 horses under him, which ought, in fact, to have been the command of a lieutenant-general of artillery.†

* Minutes of Evidence, Report of the Commissioners for Inquiry into Naval and Military Promotion and Retirement, p. 48.

† Remarks on the Organization of the Corps of Artillery in the British Service, pp. 149, 150. As this pamphlet will be quoted in the following pages, it may be as well to mention that it appears (p. xiv. of

That Sir George Wood, although the oldest officer of the whole army in the field, should still have been only a lieutenant-colonel of artillery, must obviously have been owing to some radical defect in the system; otherwise, the numerous casualties and augmentations incident to a protracted war must long before have raised him to a higher grade in this important service.

under a regimental colonel, who was the oldest officer in the field.

Before touching on the question of amelioration, it is desirable to revert very briefly to the past state of the Royal Artillery, extracted from Kane's List,* and also from "Memoirs of the Royal Regiment of Artillery, commencing 1743," in the Regimental Library at Woolwich.† In 1710 the artillery service consisted of only three large companies and twenty officers, viz., a colonel-commandant, a second colonel, two lieutenant-colonels, a major, three captains, three first lieutenants, three second lieutenants, six lieutenant-fireworkers, an adjutant, a quartermaster, and a brigade-major. Early in 1747, three companies were raised, and added to the regiment, now consisting of thirteen companies.

Strength of the artillery from 1710 to 1793.

the Report of the Commissioners of Naval and Military Inquiry, 1840) to have been written by Col. Sir Augustus Fraser, K.C.B., which the writer is able to confirm from another and no less decided authority.

* List of the Officers of the Royal Regiment of Artillery, by Lieut. John Kane, of the Royal Invalid Artillery. Greenwich, 1815, p. 78.

† This book is a MS., apparently drawn up by Col. Forbes Macbean, of the Royal Artillery, and stated in a note to have been deposited by him in 1789.

Distribution of the artillery in 1747.

Their disposition was as follows: five in Minorca, Gibraltar, Newfoundland, Louisburg, and in Scotland; five with the army in Brabant; one embarked in an expedition commanded by Admiral Boscawen, which failed at Pondicherry, in the East Indies; and two at Woolwich, having a detachment of four officers and seventy non-commissioned officers and men, at Bergen-op-zoom.

The three companies added in 1747 were reduced at the peace of Aix-la-Chapelle, in 1748.

The equipment of the field artillery with the army in Flanders, from 1742 to 1748, is stated to have been as follows:—

British field artillery employed in Flanders.

1742 and 1743. The army consisted of 19 battalions, and 19 squadrons, with 24 heavy 3-pounders.

1744. 22 battalions and 29 squadrons, with 10 heavy 6-pounders, 30 heavy 3-pounders, and 4 8-inch howitzers.

1745. 27 battalions and 26 squadrons; 10 heavy 6-pounders, 27 heavy 3-pounders, 6 1½-pounders, and 4 8-inch howitzers.

1746. 7 battalions and 9 squadrons, with 14 heavy 3-pounders.

1747. 14 battalions and 14 squadrons, with 6 heavy 12-pounders, 6 heavy 9-pounders, 14 heavy and 12 light 6-pounders, 14 heavy 3-pounders, 2 8-inch howitzers, and 6 royal mortars.

1748. 22 battalions and 14 squadrons; 6 heavy 12-pounders, 6 heavy 9-pounders, 14 heavy and 44 light 6-pounders, 2 8-inch howitzers, and 6 royal mortars.

This abstract of the troops and artillery in Flanders during the war of 1741 is given in order to show, by the quick progress and augmentation of this arm in proportion to the number of troops employed during that period, the growing sense of the importance of artillery.

Increased importance and augmentation of the artillery.

1749. During the Peace the 10 companies of artillery were stationed as follows:—In Minorca, Gibraltar, Newfoundland, and Scotland, one company at each. The company previously at Louisburg was, on the giving back of that fortress to the French in 1749, sent to Nova Scotia as a new colony. The remaining 5 companies were stationed at Woolwich, one being generally at Greenwich for the ease of quarters.

1755. In March, 4 companies with an additional major, were added to the regiment, now consisting of 14 companies. These 4 companies embarked immediately for the East Indies. In October, 2 more companies were added, making 16 companies.

1756. In April, 2 more companies raised, making 18 companies, and in May, 1 more company (Miners), making 19 companies. In June, 1756, "a great train of artillery with 5 companies of the regiment, and the company of Miners encamped at Byfield, in Surrey, till October. It consisted of 11 light 24-pounders; 14 light 12-pounders; 20 light 6-pounders; 6 light 3-pounders; 6 royal mortars: total, 57 pieces and 10 pontoons. The Duke of Marlborough, then Master-General, marched at the head of this train.

The Royal Artillery formed two battalions.

1757. In April, 4 companies added to the regiment, which now consisted of 23 companies, (exclusive of the Cadet company). On the 1st of August, the regiment was divided into 2 battalions, 12 companies in each, including the Cadet company.

1758. In April, 2 more companies added, 1 to each battalion.

1759. January 1, 4 companies raised, and added to the regiment, which now consisted of 30 companies, Cadets included. Three companies were with the Allied Army in Germany. The British artillery in the campaign of this year consisted of 10 medium and 6 light 12-pounders, 18 light 6-pounders, and 6 royal howitzers. In November, the regiment was formed into 3 battalions, 10 companies in each.

1760. The detachment of artillery, with the Allied Army in Germany, augmented to 5 companies, and the guns consisted of—

Artillery in the field in 1760.

8	heavy 12-pounders.
10	medium ,,
6	light ,,
30	light 6-pounders.
3	8-inch howitzers.
6	royal mortars.
Total 63	pieces.

1761. Two companies added to the regiment.

1762. Two companies of artillery embarked for Portugal to serve with 6 battalions, and a regiment of light dragoons under Lord Tyrawley, for the defence of that kingdom. The guns were—

Artillery employed in Portugal and Germany.

2 heavy 12-pounders.
6 medium „
6 medium 6-pounders.
12 light „
2 8-inch howitzers.
4 royal howitzers.

Total 32 pieces.

The guns of the British artillery in Germany in this year were—

8 heavy 12-pounders.
6 medium „
4 light „
24 heavy 6-pounders.
34 light „
8 8-inch howitzers.
4 royal howitzers.

Total 88 pieces.

1763. On the conclusion of the war, the 3 battalions, consisting of 30 companies, and a company of Cadets, with their respective field-officers and staff, were kept up; but the establishment of the companies was reduced.

During the Peace, 1 battalion was stationed in North America, distributed at the several stations in that continent. Another battalion in the garrisons of Minorca and Gibraltar; and one remained in Great Britain, which relieved the others.

Artillery during the peace of 1762.

The artillery formed into battalions.

1771. The regiment on the 1st of January of this year was formed into 4 battalions, each consisting of 8 marching and 2 invalid companies, with an aggregate of 199 officers, exclusive of the Royal Irish Artillery.

1779. The 4 battalions made to consist each of 10 marching companies. The 8 companies of invalids, with 2 odd companies, were formed into an invalid battalion. The establishment of each of the 40 marching companies was 1 captain, 1 captain-lieutenant, 2 first and 2 second lieutenants, 4 sergeants, 4 corporals, 9 bombadiers, 18 gunners, 69 mattrosses, 1 fifer, and 2 drummers.

Also in 1791 and 1793.

In 1791, a field officer and two strong companies were added, for service in India; and soon after the commencement of the revolutionary war in 1793, the British artillery comprised 305 officers, namely, those belonging to four troops of horse artillery and eleven invalid companies, in addition to two companies in India, and four battalions of ten companies each in various parts of the world. The marching battalions were organized almost exclusively for garrison duties, and on taking the field in the Low Countries, towards the middle of that year, drivers and horses were hired by contract for a limited period.

Defective state of the British field artillery in 1793.

The horses of these guns had the serious disadvantage of being placed one after the other in single draught, the Flemish drivers, who were on foot, having generally three horses each, which were managed by means of a long whip.* As may readily be imagined, such an imperfect equipment effectually clogged the movements of the 30,000 men who, under the Duke of York,† were opposed, in 1793 and 1794, to about the same number of the French. The latter were successively commanded by Houchard, Jourdan, and Pichegru, whose operations were supported by a powerful artillery, now become so manageable, that it held from henceforth a prominent place in the numerous victories to which it so largely contributed.

The French field artillery numerous and efficient in 1793.

The loss of part of our cumbrous material, and the example of an energetic enemy, during the campaigns in the Low Countries, led to a great, and to what might have been an effectual amelioration in our field service. But instead of making the drivers who were added a part of the companies, as had recently been done in raising the horse brigade, the Duke of Richmond, then Master-General of the Ordnance, organized it as a separate service, which in 1797 had five overgrown troops, each consisting of 275 non-commissioned officers and men, under a captain-commissary and a lieutenant-commissary, and seven quartermaster-commissaries.‡

A driver corps raised in England, consisting of five large troops.

* Remarks on the Organization of the Corps of Artillery in the British Service, p. 50.

† Alison's History of Europe, vol. ii., pp. 141, 485, 512–515.

‡ Remarks on the Organization of the Corps of Artillery, &c., p. 71, and Kane's List of the Royal Regiment of Artillery, p. 82.

State of the British artillery in 1799.

But the author finds, from one of the officers who served at this period, that when, in 1799, the artillery took the field in Holland, there was but little improvement in consequence of this arrangement. To each brigade of the army under a major-general were only attached two light 6-pounders and one ammunition waggon, having three horses, single draught, to each piece, and a driver on foot, using a waggoner's whip.

Two-wheeled ammunition cars introduced.

On the renewal of the war in 1802, another step was, however, made towards a more efficient field artillery, by the introduction of ammunition cars, one of which, like the Irish vehicle of the same name, carried six men, and followed each gun, the limber of which accommodated the remainder of the gun detachment. Such was the equipment of a car brigade of six guns, to which a strong company of artillery was attached, with a proportion of drivers under a separate officer, who had no command whatever beyond the care of the drivers and horses.

Strength of a troop of drivers.

Augmentations were, however, made from time to time, and in 1813 and 1814 the anomalous corps of Royal Artillery drivers consisted of 12 gigantic troops, each having 7 officers, 55 non-commissioned officers, 52 artificers, &c., each troop amounting, with 500 drivers, to 614 in all.* There were, besides, 10 effective battalions of

* Kane's List of the Royal Regiment of Artillery, pp. 84, 87.

10 companies each, 1 invalid battalion, 14 troops of horse and rocket artillery, and a riding troop, in all 28,291 officers and men, and 11,600 horses, viz.,—

In 1813 the artillery service numbered 28,291 officers and men.

	Officers.	Non-commissioned Officers and Men.
Foot Artillery	757	15,314
Gentlemen Cadets . . .	200	..
Horse Brigade	81	2,567
Royal Artillery Drivers .	106	7,345
Hanoverian Artillery . .	61	1,280
Foreign Artillery . . .	22	558
	1,227	27,064
	Add Officers . .	1,227
	Total . .	28,291*

The employment of the Hanoverian and foreign artillery ceased at the conclusion of the war, and the few of the old corps of drivers retained in the service were placed under the charge of the horse artillery officers who had been reduced in 1818. This change was followed by a General Order, dated September 28th, 1821, reducing the corps of artillery drivers; and, following out what had been temporarily introduced in Spain, by attaching the drivers to particular brigades of guns, the Duke of Wellington ordered that in future every man should be enlisted as a gunner and driver, and should be liable to serve equally in either capacity.

Gunners and drivers become one service.

The reductions, however, which took place,

* Kane's List of the Royal Regiment of Artillery, pp. 84, 91.

In 1823 there were only 6,050 officers and men.

gradually diminished the artillery service from the preceding strength to the very limited number of 500 officers, 746 non-commissioned officers, and 4,804 gunners and drivers, in January 1823. The new organization, by which the drivers became an integral part of each company, was pronounced by a great artillerist, the late Major-General Sir Alexander Dickson, G.C.B., to have made 5,000 men under the new, equal to 7,000 under the old system.* It was first brought a little to the test of experience in 1826, when four companies of artillery accompanied the 5,000 men sent to Portugal under Lieutenant-General Sir William Clinton, three being attached to brigades composed of one 24-pounder howitzer and three 9-pounder guns, and the fourth attached to a ball-cartridge brigade. It would appear that this equipment was chosen in preference to the horse brigade, since the latter furnished eighty-four horses to complete this service.

Advantages of abolishing the driver corps.

Expedition to Portugal in 1826, and

Introduction of the name of field battery.

The continental term of *field battery* was now adopted in England; and towards the close of the following summer, companies were attached for exercise to four-gun batteries, and an uniform system of drill was established, by order of the Master-General, on the 22nd of December, 1831.

* Minutes of Evidence before the Commissioners on Naval and Military Promotion, &c., pp. 57, 928, 939.

Thus was laid the foundation of a great and beneficial change, which was admirably calculated to create an efficient field artillery, had there been sufficient time to instruct the officers and men, and had the former still possessed the requisite bodily activity; but, unfortunately, both of these advantages were, *and still are, wanting.*

Want of time to instruct the whole regiment.

The state of the regiment became so hopeless after the war, that openings of employment in the line were made for artillery officers, and retirements, by the sale of unattached commissions, and subsequently retirement on full pay were permitted; but that the great boons thus bestowed by the Sovereign have not remedied what has been called by a great authority (Sir Henry, now Lord Hardinge) *the vice* of the artillery service, is evident from the fact, that the senior lieutenant-colonel has already been forty-seven years, and the junior of that grade thirty-five years, in the service; the latter after serving the whole of this time (the best part of his life) as a lieutenant and captain.

State since the peace.

As the officers who live long enough to reach the rank, must be at least fifty-five years of age when they are gazetted lieutenant-colonels, most of them necessarily become so infirm before the next step, that of full colonel, is obtained,* that being then usually above sixty-three years of

Ages of the lieut.-colonels.

* At present it usually requires more than eleven years to pass through the regimental rank of lieut.-colonel.

Ages of the junior colonels.

age, they no longer possess the necessary physical strength to make a fair return to their country for the pay and allowances which they continue to draw.

With very few exceptions, therefore, the colonels of artillery are only fit for ease and retirement, and as the present lieutenant-colonels are fast approaching the same condition, such a state of things is calculated to destroy hope as well as everything like ardour, and must therefore, if not remedied, eventually compromise the remaining efficiency of the corps.

State of the artillery officers in Russia and France;

In the continental armies, the artillery officer occupies a higher place, not only in point of pay, but also in rank and consideration generally. In Russia, the officer in command of a troop or battery has a rank which nearly corresponds with our major; and, as has been seen in p. 78, a superior officer is allotted to two horse or three field batteries, which, as a matter of course, improves the positions of the other officers.

In France, the officers of this arm speedily attain the rank of captain, even in time of peace; and they have, besides, the powerful encouragement of one step in three being given, by selection, for merit, the two others being awarded by seniority, with, as has been shown, ample openings for higher employment.

In Austria, Prussia, and Sweden, the artillery

officer is distinguished in proportion to his scientific attainments, and he is, in consequence, more frequently advanced than the officers of the two other arms.

In Austria, Prussia, and Sweden.

In Great Britain alone, the superior officers of artillery are behind their cotemporaries, not even excepting the officers of marines,* and are also excluded from the general staff of the army, as well as the commands at home and abroad; and, if compared with their cotemporaries in the line, the field officers at most of the home stations have a serious disadvantage with regard to pay. The senior lieutenant-colonels of artillery, for instance, receive 18*s*. 1*d*. per diem, and the juniors of this rank, who are on major's pay, 16*s*. 11*d*.; whilst corresponding ranks in the line receive 17*s*. and 16*s*. respectively. This daily pay however, with the addition of 20*l*. non-effective allowance to the senior lieutenant-colonel and major of the line, gives to both a small pecuniary advantage over the same ranks in the artillery, independently of the 3*s*. drawn by the commanding officer of the regiment. But although a large amount of artillery stores, and other causes of responsibility, require the presence of a field officer, where there may be only two or, perhaps,

Disadvantages of British artillery officers as to rank and pay,

* The last officer of the Royal Marines who has obtained the rank of full colonel entered the service the same year as the junior colonel of artillery; but he has passed through the grade of lieut.-colonel in less than five years.

one company of artillery,* this allowance is only drawn by an artillery field officer, according to the regulations of the Board of Ordnance, when there happen to be three or more companies at the station. In many cases this regulation bears hardly, particularly in this kingdom, where only one of the six field officers who are employed receives the allowance; and one of those who are excluded from this, as well as from the 20*l.* non-effective allowance, was until now the senior lieutenant-colonel, doing duty regimentally with the army in Ireland.

and command pay at many home stations.

Length of service and pay of

Pecuniary considerations would not, however, be noticed in this place, were it not for their peculiar bearing in connexion with the greater length of time that such disadvantages exist. For instance, the senior lieutenant-colonel of artillery has been forty-seven years, and the junior, receiving lieutenant-colonel's pay, almost thirty-nine years in the service, which, as regards those receiving the bare sum of 18*s.* 1*d.* per diem, is probably almost without a parallel in any branch of the service.

the lieut.-colonels of artillery.

The periods just mentioned must show that, notwithstanding a liberal scale of retirement and an augmentation of two battalions, the state of the higher ranks in the regiment has become

* Tenth Report of Commissioners on Naval and Military Promotion, &c., p. x., 1840.

progressively worse. It has been stated * that the ages of the twenty senior colonels ranged from 58 to 66 years, and their services from 42 to 44 years. The latter are now from 48½ to 54½, and their ages from 65 to 70 years. The ages of the ten senior lieutenant-colonels averaged 58 years, and the eleven juniors 51 years,† whereas the former is now increased to 62½, and the latter to 55½ years. Moreover the service of the senior lieutenant-colonel has been increased from 41 to 47 years.

Ages and services in 1840 compared with 1851.

This being the case, it is easy to anticipate the consequence of allowing such a state of things to continue.

Several circumstances have tended to retard promotion in the artillery, and one of these presses the more heavily as it is not made to apply to the other Ordnance corps. It will be recollected that, from motives of economy, even those officers who are employed in situations connected with the artillery service, whether military or civil, hold such appointments in some measure to the disadvantage of their brother officers, on whom their duties fall, frequently for a lengthened period, without the benefit of the temporary promotion in the corps, which would be the

Continuation of employed officers on the strength of the regiment.

* Report of the Commissioners on Naval and Military Promotion, &c., p. xi., 1840.

† Ibid.

Few artillery officers seconded.

consequence were those twenty-five individuals *seconded.*

To illustrate this disadvantage, it may here be observed, that four lieutenant-colonels and eight captains, belonging to the six battalions of Royal Engineers are on the seconded list, whereas there are but three officers for the thirteen battalions of horse and foot which now compose the Royal Artillery so circumstanced, although twenty-five officers are employed in various ways, such as the permanent staff in the Arsenal, situations, and at the Royal Military Academy.

It were but waste of time here to notice the difference of rank which exists between the *most fortunate* of our officers and those of his cotemporaries in the line who shared in the same active service; but it may not be out of place briefly to notice the difference which obtains in the higher ranks compared with those of other branches, in which seniority regulates the promotion.

Partly owing to the number who are seconded, it will be found that the whole of the colonels of the sister corps have had a decided advantage.

Promotion of the Engineers.

For instance, the cotemporary of the 19th lieutenant-colonel of artillery has been a colonel of the Royal Engineers since July 1851.

But the disparity in this instance is greatly exceeded in another seniority service, that of the East India Company; the cotemporary of the officer in question, and even his junior at the Academy, having obtained in 1841 the rank of major-general in the Indian army.*

Other causes, in addition to those just mentioned, have retarded the promotion of officers in the corps:—

The British artillery is smaller, in proportion to the line, than that in the continental services.

1st. The proportion of artillery, compared with the strength of the line, is decidedly smaller in Great Britain than that which exists on the Continent.

2nd. For one portion of the British army, namely, the twenty-seven regiments serving in India, no artillery whatever is maintained; this arm being provided by the East India Company, in addition to the proportion required for their own part of the service, namely, the native regiments.

3rd. The radical defect of the organization of the Royal Artillery is the existence of too many company officers in proportion to the field officers; and it will be elsewhere seen that the varying strength in time of war of the 10 adjutants' detachments averaged for each about 300 non-commissioned officers and men, who, to this extent, were under his charge, without a due proportion of even subaltern officers.

During the war a 9-pounder brigade of artillery consisted of 170 horses, 117 non-commis-

* Major-General Monteith quitted the Academy for the India service in February, 1807, and Lieut.-Col. Otway for the Royal Artillery the 1st of July, 1807.

A brigade of artillery in war-time.

sioned officers and gunners, with 123 drivers, artificers, &c., making in all 240.* With the six guns there were five company officers, and a sixth specially attached to the drivers. The whole was under the first captain, whose rank corresponded neither with his length of service, nor with the importance of the command itself, which might properly have been that of a field officer. The rank of the second captain was proportionally lower in the scale, that officer having, in fact (except when the first captain was absent), little more than the duty of a subaltern to perform.

The second captains little required.

Disadvantages of having two captains.

The establishment of five officers to each company appears to be one of the great evils which exist in the artillery service, and it retards the advancement so seriously, that it is doubtful whether, as long as this regulation prevails, any promotion, however extensive, could enable those gentlemen to attain the rank of field officers whilst they are in the prime of life.

First proposed amelioration.

The serious evil of having had fifty company officers in each battalion when complete, with only an outlet of six in the higher ranks, namely, two colonels and four lieutenant-colonels, might be remedied in two ways, both of which would be almost without expense. The one is by means of retirements, and the other by a reconstruction of the corps; the desirableness and even the neces-

* Minutes of Evidence, in 1838, before the Commissioners on Naval and Military Promotion, &c., p, 82.

sity of which has been so far admitted in evidence before the Committee of the House of Commons, that the full complement of field officers has been allotted to eight companies,* instead of making ten the standard as heretofore.

The remedy by retirements could be effected by permitting one-half of the captains to go out, either as unattached majors, or on full pay as captains, if preferred; at the same time allowing one-third of the lieutenants to have the choice of retirement, either as unattached captains, or with the full pay of their rank. Such an arrangement would at once reduce the number of company officers to thirty for a full battalion; these would have an outlet of six superior officers, and such a change would be attended with an immediate diminution of the expense for quarters, servants, &c., with a still greater saving as the annuitants drop off. The second captain being thus done away with, like the captain-lieutenants in the line, and those in the East Indian Artillery, who were removed by Lord Hastings, the remaining lieutenants would at once pass to companies, and, subsequently, from these to lieutenant-colonelcies. On the other hand, such diminution of the officers might be somewhat too great for the men to be retained;

Proposed retirement of ninety-six captains and ninety-six lieutenants.

The result of this change.

* Minutes of Evidence before the Select Committee of the House of Commons on Army and Ordnance Expenditure, 1849, p. 23, Nos. 361, 362, 363, 364.

but by a different and judicious organization, the relative working proportions might be maintained in the field, as well as during colonial service.

Projects for ameliorating the condition of the regiment.

In addition to the efforts which have already been made to accelerate promotion by the removal of artillery officers to regiments of the line, as well as by the sale of unattached commissions, and the subsequent boon of retirements on full pay, it has been suggested that a fund should be created for the purpose of augmenting the income of those who retire: this might assist in overcoming the reluctance which is often felt by an artillery officer to turn his thoughts to a different occupation, particularly if he is to enter upon it at a time of life when, from having been long habituated to military service, he is unfitted for the change: but it is very doubtful whether the limited means of artillery officers would admit of their finding the necessary funds for this purpose. It has also been proposed, as a means of overcoming such reluctance, to give brevet rank nearly similar to that enjoyed by the Guards.

The rank enjoyed by the Guards declined by the Artillery officers.

The extension of the rank of the Guards to the artillery was approved by the Duke of York as Commander-in-Chief, and the warrant to this effect was, it is understood, actually signed by the Prince Regent in 1812. But by a singular perversity of the senior officers at the time, the boon, which would have brought such life and given such energy to the corps, was frustrated. They urged that "the regimental promotion was so slow, that officers who were only captains in the regiment, might, and probably would be major-generals in the service, and have to perform duties quite incompatible with that rank." Unfortunately these objections prevailed, and the warrant was cancelled.*

Retirements with additional rank, as

But if for the sake of efficiency the country were inclined to overlook the expense, there is no doubt that something similar to the arrangement lately adopted for the retirement of old officers of the Navy, or that of the East India Company for the same object, would give effective relief to the two Ordnance corps. In the former service it has been found that the temptation of a flag on the retired list, was not to be resisted by captains of a certain standing, whose promotion would not take place in turn for some years to come; and in the latter, the scale of retirements has worked the more beneficially, since it is not confined to any particular rank. By the regulation of December, 1836, the time of retirement reckons by service on full pay in India—viz., an officer after twenty years, retires with the rank and pay of

in India, would benefit the corps.

* Report on the Numerical Deficiency, Want of Instruction, and Inefficient Equipment of the Artillery of the British Army, by Major-General Sir Robert Gardiner, K.C.B., published in Jones's Woolwich Gazette for July, 1848.

captain; after twenty-four years with that of major; after twenty-eight years with that of lieutenant-colonel; and after thirty-two years service with the rank and pay of full colonel.*

Too many generals would be the consequence.

Doubtless the speedy attainment of the rank of major-general in this manner, would give comparatively rapid promotion to those of lower grades; but as this would be attended with the disadvantage of increasing the already extensive list of general officers in the British Army who are without employment, and perhaps cause some dissatisfaction also in other branches of the service, it is worthy of consideration, whether the desired object might not be attained by a more suitable proportion of the superior to the junior regimental officers, and, at the same time, put an end to the system of doing duty by separate companies under field officers who are strangers to the men.

It will be admitted, that a division of an army, consisting of regiments known to one another, by having been previously brigaded together, would be more efficient when in the presence of an enemy, than another division, which might be equally good, but which is formed of regiments brought together for the first time, as was the case in Holland, during the expedition of 1793, and again in 1799, when the regiments forming

* Minutes of Evidence before the Commissioners on Naval and Military Promotion, &c., pp. 37, 38.

brigades, and the generals commanding the brigades were unknown to each other.

For the same reason, if ten choice companies belonging to as many distinguished regiments were assembled under different field officers, they would for a time, form an imperfect corps: yet this sytem prevails in the artillery; for the companies both at home and abroad are assembled at stations by accident, under field officers, who, coming together, also by chance, from various battalions, must consequently be quite ignorant of the characters of the men, and usually but little acquainted with the qualifications of the officers thus assembled. For example, at one of our most important foreign stations (Gibraltar), there are five companies, namely, No. 8 of the 2nd battalion, No. 3 of the 3rd, No 4 of the 7th, No. 7 of the seventh, and No. 7 of the 9th, under a colonel of the 6th and a lieutenant-colonel of the 2nd; so that there are only two companies of the same battalion: and recently even this was not the case, the whole having been in 1848 of different battalions. The disadvantages caused by this state of things must be manifest, in carrying out the drills and duties, which, as in the case of different regiments, are too apt to vary also in various companies. It is not, however, difficult to find a remedy for the evil almost without any increase of expense.

Disadvantages of doing duty by isolated companies.

Existing defects of the regiment.

The principal defects in the existing state of the artillery service are as follow :—

1st. Slowness of promotion in the higher ranks.

2nd. A superabundance of company officers; or rather a want of a due proportion between the field and the company officers.

3rd. Frequent changes of stations and companies in consequence of having so many grades of officers.

4th. Doing duty by companies instead of battalions.

5th. Difficulty of carrying out one uniform system of duties throughout the ninety-six companies of the regiment.

6th. Want of *esprit de corps* in consequence of the existing system, and also of that unanimity which may belong to a battalion, but can hardly be maintained throughout ninety-six separate companies.

7th. The field officers are deprived of the advantage of thoroughly knowing the officers and men serving under their orders; and those employed on home commands receive, as has been shown, less pay than officers of the line of their rank and standing. The necessity of ameliorating some of these disadvantages by means of a reorganization, appears to have been felt many years back, and the following remarks, which have been recently perused by the writer, cannot fail to have weight

Remedies proposed by a distinguished officer.

as giving the opinion of a very distinguished officer, long one of the brightest ornaments of the service, the late Colonel Sir Augustus Fraser, K.C.B.

"But the chief means of accelerating the promotion, and of really improving the corps, would be by finding employment for all the superior ranks, many individuals of which, according to existing arrangements, cannot be said to do any duty.

Improvement of the regiment.

"A simple mode presents itself of doing this, and one which is equally applicable to peace or war. It is to divide the battalions into two parts, of five companies each, under the title of brigades; to place the two brigades of each battalion under the command of the junior colonel of the battalion, distributing the lieutenant-colonels equally between the brigades; and to do duty in every part of the world, not by separate and independent companies, as at present, but by battalions, brigades, or component parts of a brigade; and further to require, that all field officers of every rank, the colonels commandant and senior colonels excepted, accompany and remain with their battalions and brigades wherever they may be stationed, in the same manner as is expected of officers of other parts of the service. . . . .

"The subject has hitherto been viewed solely in the light of benefiting the service by the quicker promotion of officers. Let us now consider its probable effects in improving the efficiency of the corps.

"Let us recollect the present arrangement of the corps, that the duty is done by separate companies, seldom, it may almost be said never, of the same battalion; that the field officers can have no previous knowledge of or responsibility for the instruction of the men, whom they have perhaps never seen till the moment of having them placed under their almost nominal command; that the companies come from various

Nature of the present system.

quarters, and with very various preparations for service; that some companies have perhaps been for years in a foreign garrison, where they may have performed none but infantry duties, and in some cases may be commanded by captains who have just joined them.

Want of a gradation of command and responsibility.

"Let us compare this strange jumble (and the assemblage of the artillery in Belgium a very few weeks previous to the battle of Waterloo was very like it) with the well-disciplined and well-appointed corps which must be forthcoming, if previously, habitually, and systematically formed under the care of field officers identified with it, and responsible for its instruction and efficiency.

"No one can for a moment deny that order, system, and instruction, are necessary to success. . . . .

"By the gradations of command and responsibility here proposed, while the captains of companies would continue equally responsible as at present for the discipline and instruction of these particular companies, the several companies of a brigade would be placed under the more immediate superintendence of field officers, responsible for their respective commands to the colonel commanding the two brigades, or battalions, who would himself in his turn be responsible for the whole to his own superiors."*

To retain officers or men who are inefficient is but to incur useless expenditure; and it will be readily admitted that, whatever number of both the country can afford to pay, the ranks should bear a just proportion to one another, and be at the same time on the most efficient working footing; so as to give fair promotion to the officers, with more hope, and consequently better

* Remarks on the Organization of the Corps of Artillery in the British Service, pp. 24–28. Rowland Hunter, London, 1818.

encouragement to the men. The efficiency of the corps through such means being the immediate object, it is worthy of consideration whether some change of construction similar to that which formerly took place in India might not answer the purpose.

Reconstruction of the East India Artillery.

Unwieldy battalions and companies having an undue proportion of officers prevailed in that country up to 1818, when a remedy was sought and found by a distinguished soldier and statesman, the Marquis of Hastings, then Earl of Moira. Before his time each company consisted of ninety men with six officers, viz., a captain, a captain-lieutenant, two first-lieutenants, and two lieutenant fire-workers; and as each battalion numbered seven of these companies, and had only four field officers, viz., two lieutenant-colonels and two majors, there was only an outlet of four superior officers, including the adjutant and quartermaster, or forty-four company and staff officers. To remedy this glaring disproportion and the other disadvantages of such a formation, Lord Moira did away with two officers in each company, absorbing them by promotion chiefly. Instead of the previous large battalions, the latter henceforth consisted only of five companies, four of these having gunners as usual, and the other, with a view to providing for staff employments, officers only, namely, one captain, two first lieu-

Two officers of each company abolished.

Proportion of company to field officers.

tenants, and one second lieutenant; for whom, that is the twenty company officers in a battalion, there is an outlet of three superior officers, viz., a colonel, a lieutenant-colonel, and a major.*

Remedies proposed.

Taking the preceding as a basis for the reorganization of the Royal Artillery, it is proposed—

1st. To diminish the number of officers by dispensing with the second captains and second lieutenants; thus reducing the grades to four.

2nd. To alter the existing proportions between the ranks, so that there shall be more field officers.

3rd. To form smaller companies and battalions having field officers attached, and doing duty with them, which may be better suited than those of the present strength for foreign stations.

4th. To divide the service into field, and heavy or garrison artillery, as in the continental armies.

Formation of the battery and company.

It is assumed as the basis of this change that there should be a subaltern for two, and a captain for four guns, so that a field battery might either consist, as in the Austrian, Russian, and Prussian armies of eight guns, or merely the present battery of four guns. The latter, or rather the company attached to it, has been chosen as the unit, on account of the numerous detached services which fall to the lot of the British army.

The new company in time of peace and war.

In conformity with this idea, it is proposed that a company, in time of peace, shall consist of

* In the Golundauz battalions the proportion is still more favourable, there being five superior officers to eight captains.

one captain and two first lieutenants (one to each demi-battery), nine non-commissioned officers, thirty-four gunners, twenty-four drivers, four artificers, and one trumpeter, or in all seventy-two non-commissioned officers and men, with fifty horses, which in time of war would be augmented to eleven non-commissioned officers and eighty-nine gunners, drivers, &c., or one hundred in all, having for a 6-pounder battery, with two spare ammunition waggons, eighty or eighty-five horses, and for a 9-pounder battery, also with two spare waggons, ninety-five or one hundred horses six such companies, or 452 non-commissioned officers and men in the former case, and 600 in the latter, to constitute a battalion, and to do duty invariably as such, at home as well as abroad, whether in time of peace or of war.

Six companies to a battalion.

A battalion on this scale will be found suitable for most of the foreign stations; and, when taking the field, it would, besides having two companies with the reserve, be sufficient for sixteen guns. This number would be only a fair proportion for a division of the army, or it might occasionally, though inadequately, suffice for two divisions. In the latter case there would be a lieutenant-colonel and two batteries (eight guns) to each division, or a battery to each brigade, having the third field officer and two companies, or one-third, in reserve. The latter may seem a large proportion, but it should be borne in mind that,

Sixteen guns to a division as a full complement; or, on a reduced scale, one battery to a brigade, and a lieut.-colonel for two batteries.

One-third in reserve, and its duties.

Duties of the reserve.

besides keeping complete those field batteries which are more actively employed, many important duties devolve upon this portion of the service, such as the battering artillery, guns of position, and reserve ammunition, both for the batteries and small arms.

—	A Troop of Horse Artillery, with Two light 6-pounders and Two 12-pounder Howitzers, and Two Ammunition Waggons. Peace Establishment.							A Troop of Horse Artillery, with Two 6-pounders and Two 12-pounder Howitzers, and Four Ammunition Waggons. In time of War.						
	Guns and Howitzers.	Carriages.	Officers.	Non-commissioned Officers and Men.	Artificers.	Horses.	Rounds of Ammunition.	Guns and Howitzers.	Carriages.	Officers.	Non-commissioned Officers and Men.	Artificers.	Horses.	Rounds of Ammunition.
Guns	2	2	..	..	..	8	120	2	2	..	..	..	12	194
Howitzers	2	2	..	..	..	8	86	2	2	..	..	..	12	136
Rocket Tubes	..	1	..	..	..	4	..	..	1	..	..	..	6	..
Ammunition Waggons	..	..	..	..	..	2	..	..	4	..	..	..	24	..
Forge Waggons	..	1	..	..	..	4	..	..	1	..	..	..	6	..
Spare and other Carriages	..	..	..	..	..	..	..	..	2	..	..	..	8	..
Captains	..	..	1	..	..	..	..	..	..	1	..	..	..	..
Lieutenants	..	..	2	..	..	..	..	..	..	2	..	..	..	..
Staff-Sergeants	..	..	..	1	..	1	..	..	..	..	1	..	}	..
Sergeants	..	..	..	2	..	2	..	..	..	..	3	..	11	..
Corporals	..	..	..	3	..	3	..	..	..	..	3	..	}	..
Bombadiers	..	..	..	3	..	3	..	..	..	..	4	..	}	..
Gunners	..	..	..	44	..	24	..	..	..	..	54	..	..	..
Drivers	..	..	..	20	..	..	..	..	..	..	42	..	24	..
Trumpeters	..	..	..	1	..	1	..	..	..	..	2	..	2	..
Farriers	..	..	..	..	1	1	..	..	..	..	..	2	}	..
Car-smiths	..	..	..	..	1	1	..	..	..	..	..	1	}	..
Shoeing-smiths	..	..	..	..	1	1	..	..	..	..	..	1	6	..
Collar-makers	..	..	..	..	1	1	..	..	..	..	..	1	}	..
Wheelers	..	..	..	..	1	1	..	..	..	..	..	1	}	..
Baggage and Spare Horses	..	..	..	..	..	..	..	..	..	..	..	..	4	..
Total	4	6	3	74	5	65	206	4	12	3	109	6	115	330

The troop on the Peace Establishment has two ammunition waggons attached, with, including the limbers, 120 rounds for the guns and 86 rounds for the howitzers.

On the War Establishment there are four ammunition waggons, giving 194 rounds for the guns and 136 rounds for the howitzets, with a further supply with the reserve.

The following table gives the details of the proposed troop of horse artillery in times of peace and war; also of a 6-pounder battery during peace, and of a 9-pounder battery, which is supposed to be substituted for the latter in case of war:— Details of troop and battery.

Field Battery, with Three 6-pounders and One 12-pounder Howitzer. In time of Peace.							Field Battery of Three 9-pounders and One 24-pounder Howitzer, with Two additional Ammunition Waggons. In time of War.							
Guns and Howitzers.	Carriages.	Officers.	Non-commissioned Officers and Men.	Artificers.	Horses.	Rounds of Ammunition.	Guns and Howitzers.	Carriages.	Officers.	Non-commissioned Officers and Men.	Artificers.	Horses.	Rounds of Ammunition.	—
3	3	..	..	..	12	194	3	3	..	..	..	18	148	Guns.
1	1	..	..	..	4	136	1	1	..	..	..	6	144	Howitzers.
..	..	..	..	..	..	..	..	..	..	..	..	..	..	Rocket Tubes.
..	4	..	..	..	16	..	..	6	..	..	..	36	..	Ammunition Waggons.
..	1	..	..	..	4	..	..	1	..	..	..	6	..	Forge Waggons.
..	..	..	..	..	..	..	..	2	..	..	..	8	..	Spare and other Carriages.
..	..	1	..	..	..	..	..	..	1	..	..	1	..	Captains.
..	..	2	..	..	4	..	..	..	2	..	..	2	..	Lieutenants.
..	..	..	1	..	4 (Staff-Sergeants to Bombadiers)	..	..	..	..	2	..	6 (Staff-Sergeants to Bombadiers)	..	Staff-Sergeants.
..	..	..	2	..		..	..	..	..	3	..		..	Sergeants.
..	..	..	3	..		..	..	..	..	3	..		..	Corporals.
..	..	..	3	..		..	..	..	..	3	..		..	Bombadiers.
..	..	..	34	..	..	..	..	..	..	45	..	..	..	Gunners.
..	..	..	24	..	..	..	..	..	..	36	..	..	..	Drivers.
..	..	..	1	..	1	..	..	..	..	2	..	2	..	Trumpeters.
..	..	..	..	1	..	..	..	..	..	..	2	..	..	Farriers.
..	..	..	..	..	..	..	..	..	..	..	1	..	..	Car-smiths.
..	..	..	..	1	..	..	..	..	..	..	1	..	..	Shoeing-smiths.
..	..	..	..	1	..	..	..	..	..	..	1	..	..	Collar-makers.
..	..	..	..	1	..	..	..	..	..	..	1	..	..	Wheelers.
..	..	..	..	..	..	..	..	..	..	..	..	10	..	Baggage and Spare Horses.
4	9	3	68	4	45	330	4	13	3	94	6	95	292	Total.

The 6-pounder field battery on the Peace Establishment has, with the waggons, 194 rounds for the guns and 136 rounds for the howitzer. In War the 6-pounder battery is supposed to have 80 or 85 horses. Two waggons are added on the War Establishment, which will give the 148 rounds for the guns, then supposed to be 9-pounders, and 144 rounds for the howitzer. A further supply is supposed to be carried with the reserve.

Eight-gun batteries proposed by the Duke of Wellington.

It should be observed that the Duke of Wellington preferred larger batteries than those used in the Peninsula; and in a Minute sent to a committee of officers assembled at Woolwich, in August, 1820, to report upon the field equipments of the artillery, his Grace observed, that the batteries should consist of eight pieces, that is, of six guns and two howitzers, and that a company of 110 men would be equal to the management of this number.

Opinions of the committee of field officers.

The committee is understood to have stated, in reply, that a company of artillery was quite equal to the management of eight guns, and that it would be easy to make this allotment; but that the smaller number of six guns permitted the employment of the surplus men with the reserve guns, ammunition and musket-ball cartridges, escorts, &c.

Sir William Robe, one of the committee, whose brilliant and extensive services are well known, likewise thought (as we have seen that Sir Robert Gardiner did also) that an eight-gun formation was advisable; but the supposed expense outweighed other considerations: the incorporation of the drivers, however, was an important part of the recommendations of this committee.

Four-gun batteries adopted since the peace.

As the batteries were subsequently fixed at four pieces of ordnance, this number was adopted in the project of the writer, as being suited for the ordinary wants of Great Britain; but if the eight guns should be established, four guns will become a demi-battery, since the construction will obviously be the same whether the eight guns are called a battery or a double battery, there being in either case seven officers to eight guns, and three officers to four guns, if the latter are detached.

Comparison between four and six gun batteries.

The change from five officers with six guns to seven officers with eight guns will make a slight difference in favour of the latter, which will be apparent if an equal number, say twenty-four pieces, be compared in each case. Four six-gun batteries in war-time would be twenty-four guns, with, including the field officers, twenty-two officers; and three double or six single four-gun batteries would equally number twenty-four pieces, and have twenty-four officers in all, viz., one colonel, three lieutenant-colonels, six captains, twelve lieutenants, one adjutant, and one quartermaster. Moreover, it is presumed that the reserve ammunition, and the other important objects alluded to by the committee, would be better provided for by the reserve just proposed than by detaching extra gunners from the battery for this purpose.

The present strength would form twenty-four grand divisions of 486 men.

The regiment as it stands would form twenty-four battalions or brigades of artillery, each consisting of—

Details of a battalion.

1 colonel commandant.
1 colonel-en-second.
3 lieutenant-colonels.
6 captains.
12 first lieutenants.
1 adjutant.
1 quarter-master and acting paymaster.
452 non-commissioned officers and gunners.

This would give 10,416 non-commissioned officers and men, or the present strength of the regiment; and, in case of war, it might, including the Horse Artillery, be increased to about 15,044 men, without any increase of officers, the companies being in that case augmented to 100 non-commissioned officers and men. The latter number would preserve, as nearly as possible, the same relative proportion of officers and men as was adopted in 1838, and in fact rather more officers than there were on service during the Peninsular war; also rather more than in the East India Company's artillery, with a still greater number of officers than is allotted to a certain aggregate of men by continental nations. The same proportion obtains in garrison service; for having, according to the proposed plan, as many officers at the various foreign stations as there are at present, though of different grades, there would be the same means of obtaining leave of absence as is permitted by the Queen's Regulations; and a skeleton company recruiting at

The proposed scale would give nearly the same number of officers at foreign stations.

home, which is advisable, would greatly facilitate this object. It will be seen that the increase of expense caused by promotions in the several ranks will be only about 859*l.* 13*s.* 3*d.* for the regiment at large.—(See Schedules 1, 2, and 7.)

How carried out.

Should the preceding or some similar organization be considered likely to work well in practice, the change may be effected without difficulty or confusion, and at the same time would lead to very little expense incurred in consequence of the interchange of officers and men.

The companies taken by fours at home and abroad; see Schedules Nos. 3 and 4;

The ninety-six companies of which the regiment is now composed are supposed to be taken by fours, as they stand on the rolster, at home and abroad, as far as the proximity of the stations will admit,* the field officers being attached to these portions in such succession as will give to each individual, as nearly as possible, the same tour of service; namely, by posting those next for service to the companies about to embark, and those who have nearly completed their foreign service to the companies about to come home, and so on. Each section of four companies is then organized as a battalion of six companies, the men and officers continuing where they may be, so far as there may be vacancies.

and field officers posted to the new battalions.

By this arrangement all the field officers, six

* See Schedules Nos. 3 and 4.

The field officers, six captains, and twelve lieutenants would remain where they are.

of the captains, twelve lieutenants, and nearly the whole of the non-commissioned officers and gunners, would continue where they now are, and occasionally the two other captains would also be retained by promotion; and the few moves that might still be necessary, after completing the several ranks in each battalion, could be effected equitably, as regards position, by posting officers to the nearest battalions where there happen to be vacancies.

The twelve additional colonels-commandant receive pay as major-generals. Twelve lieut.-colonels on majors' pay, and 144 first lieutenants on second lieutenants' pay.

By Schedule No. 2, it will be seen that of the twelve additional colonels-commandant are not supposed to cause any additional expense, being as presumed already paid as major-generals.*

The twelve colonels-en-second, formerly proposed to be added on lieutenant-colonels' pay, being now omitted, this grade remains as it was, with the exception of the reduction of one; but, as heretofore, the junior lieutenant-colonel of each battalion is paid as major, and lieutenants of the second class (half the number) as second lieutenants. The additional officers proposed, viz., one to each of the battalions stationed abroad, would have given an important outlet where it is most wanted; but possibly the same object might be accomplished without any increase of number

* Excepting in such a case as the recent small promotion, there will usually be somewhere about thirteen major generals, consequently no expense would attend their being on the strength of the corps.

or expense, by bestowing the brevet rank of colonel on the thirteen senior lieutenant-colonels of the regiment.

Lieut.-colonels on majors' pay.

Sir Hew Ross seemed to apprehend * that this proposal with regard to the pay would be a "depôt of discontent;" but an additional grade of rank has not hitherto been considered a cause of dissatisfaction to officers, nor has it, in fact, been so to the majors of the artillery, on whom Lord Anglesey conferred a great boon in 1827, by promoting them to lieutenant-colonelcies, without increase of pay.

The proposed change in the horse brigade would (without diminishing the pay of the officers) decrease the annual expense of this part of the service by 6,372*l.* 14*s.* 4½*d.*, while the additional captains and superior officers necessary for a fresh organization of the corps would only occasion an increase, as already mentioned, of 859*l.* 13*s.* 3*d.*

Four companies, equal to a new battalion, are relieved every year.

Let us now consider how such an alteration would work at foreign stations, for at home there cannot be any difficulty.

Inadequate proportion of artillery at home and abroad.

The inadequate proportion of artillery to the other arms on home service, and its deficiency abroad, is particularly noticed by the Committee in their Report,† where it is stated, with respect to the field service, that there is but one gun to

* No. 6062, Committee on Ordnance Expenditure, 1849.

† Army and Ordnance Expenditure, 1849, pp. ix. x.; also, Nos. 312, 313, 314, 867, 868, 869.

750 men of the army at large; and "that the whole of our colonies require additional artillery."* Adverting to the colonies it will probably be admitted, as well as felt, that forty-eight instead of forty-one† companies of artillery, as at present, are required for the duties in our numerous colonies and garrisons. These will be represented by seventy-two companies, or twelve new battalions, which might be thus distributed:

Stations, &c. of the new battalions abroad.

four in the Mediterranean, two in Canada, one in Halifax, New Brunswick, and Newfoundland, one in Jamaica, and the remaining four in the East and West Indies.‡

With the exception of an acting paymaster to these battalions (whose duty might be taken by the quartermaster), and another colonel, no change would be necessary beyond a small increase of gunners and drivers, to meet particular circumstances; since the companies and battalions are adapted either for a moderate increase or diminution of men, without any change in the number of officers. §

Home stations of the new battalions.

The remaining forty-eight companies of the regiment would form twelve of the proposed battalions, which may be thus distributed: four battalions at Woolwich; two and a half in Ire-

* Army and Ordnance Expenditure, 1849, pp. ix., x.; also Nos. 312, 313, 314, 867, 868, 869.

† Ibid., No. 582.

‡ Schedule No. 4.

§ See Schedules Nos. 2 and 5.

land; one at Devonport; one at Portsmouth; one in Dover, Guernsey, Jersey, and Alderney; one at Chatham; and one and a half distributed in Scotland, Weedon, Hull, Manchester, Chester, &c.*

With regard to the relative proportion of officers, there are, agreeably to the present arrangement, for

48 companies at home	240	officers
Field officers	27	
	267	
With 12 battalions of 24 officers each there would be	288	
Giving in favour of the new organization	21	officers

Since it is well known that one large machine works better and at less expense than several smaller ones giving the same power, the change from a company to a battalion system will be unattended with difficulty, whilst it will, on the other hand, lead to the following advantages:— The proposed battalion system.

1st. The character of the officers would be mutually well known, and those of the men would, at the same time, be perfectly known to all.

2nd. The senior officer would be responsible for the efficiency of his battalion.

3rd. A uniform system of drills, duties, and interior economy would prevail, and could be carried out with more ease and certainty than is practicable at present.

* See Schedule No. 3.

4th. The battalions would, like separate regiments, give rise to an esprit de corps, with, as the natural consequence, a greater degree of unanimity amongst the officers than can exist under present circumstances.

5th. Each officer would be liable to only two changes of battalions during his lifetime; namely, once when promoted to a company, and once when promoted to the rank of a field officer.

6th. The public accounts would be greatly simplified, those of each battalion being audited together, instead of by single companies.

7th. It would be the means of facilitating the reliefs of the foreign as well as the home stations, which otherwise might not be effected in case of hostilities. Of late years, the utmost regularity and fairness of turn have prevailed in these matters, with such advantages to the service as can only be fully appreciated by those who recollect the former want of system, and its mischievous consequences.

During the war, the companies remained abroad, and

As there were, comparatively, but few companies at head-quarters during the war, the duties were chiefly taken by a large detachment under the adjutant of each battalion, from which the companies stationed elsewhere were kept up to their strength both at home and abroad; the pressure of the service, for the most part, caused the companies, particularly those at foreign stations, to remain wherever they happened to be for a very lengthened period. It is true that promotion or exchange occasionally removed officers and non-commissioned officers to other stations, if spared by the climate, but with these exceptions, in which the men had no share, death was the only prospect of curtailing the banishment. Companies sometimes returned home after the peace, without bringing back a single man who had gone out originally with them.

were filled up from the adjutants' detachments at home.

A well-known case of this kind may be mentioned which occurred in a healthy climate: No. 1 of the 2nd battalion, then Major George Fead's company, embarked for Newfoundland in 1789, and was brought back by his son, as its captain, in 1819, with only one man who had disembarked with it in the former year, and he was born on the outward passage. Another, Knox's No. 2 company of the 2nd battalion, embarked for Minorca in 1781, proceeded to Gibraltar in 1790, and returned in 1819, with only one individual of the original number, who went out an infant in arms, and returned an old soldier and a non-commissioned officer.

Fead's company remained thirty years at Newfoundland.

Other instances of lengthened banishment may also be given. No. 3 of the 2nd battalion, Campbell's company,* proceeded to Gibraltar, in 1790,

Campbell's company was thirty-eight years abroad.

* Returns of the Distribution of the several Battalions of the Royal Artillery, ordered by the House of Commons to be printed July 11, 1817, and corrected by one of the Adjutants who subsequently quitted the corps.

and returned in 1814, after serving in various places for a period of 24 years.

Caddy's thirty years.

No. 3 of the 4th battalion, Major Caddy's in 1817,* went to Canada in 1790, and returned to Woolwich in 1820, after thirty years' foreign service.

Godby's twenty-six years.

No. 6 of the 2nd battalion, Major Godby's company in 1817,† embarked for Jamaica in 1789, where, with the exception of a short service at St. Domingo, in 1795, it remained till 1815, having been twenty-six years in the tropics.

Campbell's twenty-six years.

No. 7 of the 2nd battalion, afterwards Major D. Campbell's,‡ and the writer its second captain, was also twenty-six years in Jamaica, *i. e.*, from 1789 to 1815.

Onslow's twenty-four years.

Onslow's company,§ No. 5 of the 1st battalion, went to St. Domingo in 1795, and to Jamaica in 1798, where it remained until 1817, thus completing twenty-four years' tropical service.

Cleave's twenty-three, and Light's twenty years.

Cleave's, No. 3 of the 7th battalion, completed twenty-three years in the West Indies, and Light's, No. 2 of the 1st battalion, twenty-two years.‖ Younghusband's, No. 10 of 4th battalion, was nineteen years in the West Indies; Bettesworth's, No. 6 of the 7th battalion, the same number of years in that part of the world. Power's, No. 5, was eighteen years, and Clements', No. 7 of the

* Returns of the Distribution of the Royal Artillery, July 11, 1817.

† Ibid. ‡ Ibid.

§ Ibid. ‖ Ibid.

same battalion, twenty-one years there also; Armstrong's, No. 4 of the 1st battalion, fourteen years; and Skinner's, No. 6 of the 6th battalion, eighteen years in Ceylon.

The periods at more desirable foreign stations were equally protracted. No. 6 of the 4th battalion, Sinclair's company, in 1817,* embarked for Canada in 1790, and returned to Europe in 1815, after twenty-four years' foreign service. Kirby's,† No. 6 of the 1st battalion, was twenty-three years at Gibraltar; and Skyring's, No. 10 of the 8th battalion, seventeen years at the same place: Story's, No. 8 of the 2nd battalion, twenty-six years at Halifax and Canada; Pierce's, No. 7 of the 8th battalion, seventeen years at Malta; Wilgress's, No. 3 of the 5th battalion, seventeen years at the Cape. Other instances abroad,

As the natural consequence of such prolonged service abroad, other companies remained for lengthened periods at home, changes of station being confined chiefly to those portions of the regiment which were sent on active service. Amongst numerous instances of companies remaining in Great Britain, two may be mentioned; the one, Major Desbrisy's company, which remained at Hull from 1803 to 1815: the other concerned the writer, who for most of the time as well as at home.

* Returns of the Distribution of the Royal Artillery, July 11, 1817.
† Ibid.

was one of its subalterns. No. 3 of the 6th battalion, which after being six years at Portsmouth, in the height of the war, was sent to Guernsey in 1808, and it cannot be forgotten, even at this distance of time, that while the several regiments of the line, also forming part of that garrison, were replaced one after another, and sent to the Peninsula, the company of artillery continued stationary until the peace of 1814, when it was sent to relieve one of the companies already mentioned as having been so long in Jamaica.

Colonial service confined to one portion of the corps,

Such disadvantages to the service as the preceding, which are too glaring to require any explanation, were held to have been altogether unavoidable. But although the more pressing demands of hostilities caused serious difficulties in carrying out reliefs, the means undoubtedly existed of interchanging from time to time the companies at home, and of relieving at certain intervals, those serving abroad also; and thus avoiding a state of things which, in point of fact, condemned a large portion of artillery to colonial banishment, whilst service in the field fell almost exclusively to one more fortunate section of the regiment.

and field service to another section of the corps.

Usually three-fourths of a company employed abroad, and

Usually, only those companies employed in the field had the full complement, or nearly so, of 120 gunners; the remainder, say nine-tenths of the regiment, having a certain proportion of this number with the adjutant's detachment at head-quarters. As these companies remained at certain stations, and were completed by drafts when necessary, it is obvious that during thirty-eight years in the Mediterranean, twenty-six in Jamaica, and other protracted periods elsewhere, there must have been such a number of death vacancies filled up as would, in all probability, have formed a second, and in some instances even a third company for its relief; consequently, the surviving officers did duty with much more than the ordinary proportion of men.

filled up from home, without being relieved.

At this distance of time, the writer has not sufficient means to give the exact numbers; but it is not difficult to show the general effect on the regiment at large, as it comes within the knowledge of the writer that two of the battalions were nearly thus distributed towards the close of 1813. Let us take one of the number as a specimen:—

Strength, abroad and at home, of one of the battalions of artillery.

		Gunners.	Non-com. Officers.
At Woolwich . . .	1 company. .	94	17
In Portugal, Spain, and Holland	4 companies .	367	60
At the Cape, the Isle of France, Malta, Halifax, and Barbadoes .	1 company each	469	60
The adjutant's detachment at Woolwich		270	33
		1,200	170

Probable aggregate strength of the adjutants' detachments in 1813.

Presuming the other battalions to have been nearly the same as the one in question, which had its full proportion on both colonial and field service, there must have been about 3,030 non-commissioned officers and men under the several adjutants, who, strange to say, had little more than the assistance of the non-commissioned officers in carrying out the instruction of this body of men, the officers under them being almost entirely confined to a few second lieutenants who remained with their battalions for a time to learn their duty. Should the preceding statement appear too great, a smaller scale based on the following may be taken.

Nominal establishment of a company of artillery at various periods.

It appears that a company of artillery consisted—

—	Officers.	Non.-Commissioned Officers.	Gunners.	Total.
* In 1794 of	5	15	111	131
1798	5	15	104	124
1802	5	13	86	104 {time of peace.
1803	5	16	103	124
1804	5	16	123	144
1805	5	17	122	124
1808	5	17	126	148
1808 to 1816	5	17	120	142

Aggregate strength of the ten adjutants' detachments about 3,100 non-commissioned officers and gunners;

The average of the whole of the years for the preceding period gives 16 non-commissioned officers and 127 gunners for each company; but taking the lesser average between 1808 and 1816,

* Kane's List of the Royal Regiment of Artillery, pp. 85, 86, 91.

namely, 120 gunners, and deducting therefrom 93 gunners, being the average of the companies belonging to the battalion already noticed, there will be 27 gunners, or, including the 4 non-commissioned officers, 31 for each company of the regiment, and giving about 3,100 non-commissioned officers and gunners for the 10 adjutants' detachments at Woolwich. This, considering that the number of recruits constantly placed the regiment above the supposed establishment, probably does not exceed the actual number of the great reserve at Woolwich, which in this case might have been formed into 3 battalions, or 30 companies, of the usual effective strength. But if we even suppose the 10 adjutants' detachments to have amounted to only half this number, the 1,550 would still have been equal to 15 companies; and by simply adding officers (which in time of war would not have been considered a serious expense), a battalion and a half would have been made available for reliefs at home, in the first instance, and afterwards for those abroad. The latter, it is true, would not have been so frequent as at present; but still a great benefit might have been obtained, if, instead of sending drafts to fill up the old companies, the additional gunners had been sent out at certain intervals as complete companies, to relieve those which had been most reduced by the effects of climate.

3,100 non-commissioned officers and gunners, and

about fifteen companies available for reliefs.

CHAPTER VI.

BRITISH HORSE AND FIELD ARTILLERY COMPARED.

The Company System causes want of Uniformity.—Want of Esprit de Corps.—Proposed Battalion System, and its advantages.—More Officers to be employed at Foreign and Home Stations than at present.—The Artillery Service requires more Field Officers than the other arms.—The Stores and Equipments should be in the charge of Field Officers of Artillery.—Proposed Separation into Horse, Field, and Garrison Artillery.—A regular Pontoon Train to be organized.—Duties of the Artillery Soldier.—Advantages of a Separation of the Garrison from the other part of the Service.—Eighteen Battalions to be allotted for Garrison, and six for Field Service.—Officers for the Field Artillery to be selected.—A Depôt of Drivers to be formed as a Reserve.—Supposed Falling-off of the Artillery Recruits.—Objections to entering the Artillery compared with past times.—Change of the Class of Men who enlist.—Opinion of the Duke of Wellington regarding Recruits.—Decrease of Men who can Read and Write.—Necessity, according to the present plan, of Seconding Officers who are employed.—No increase of Officers required for an establishment of 15,044 Non-commissioned Officers and Men.—The Author did not propose to discontinue the Horse Artillery.—Necessity, in his opinion, of this arm.—Distinction between the Horse and the ordinary Field Artillery.—Both kinds strongly advocated.—Efficiency of Field Batteries in 1818.—Instance given, and opinion of a French Artillery Officer on this subject.—Field Batteries might be converted into Troops of Horse Artillery.—Particular objects to be attained by Horse Artillery, when armed with Light Guns only.—Opinion given by Lieut.-General Sir Robert Gardiner on this subject.—The Nine-pounder Troop of Horse Artillery could not accompany General Gilbert's Cavalry in pursuit of the Sikhs.—Necessity of maintaining some Horse Artillery, even in time of Peace.—Evidence proposed to be given to the Committee of the House of Commons on the Field Batteries.—Seven Troops of Horse Artillery would be an adequate proportion for the British Army.—Twelve-pounder Guns and Rockets to be part of the Field Service, in addition to the Nine-pounders.—Guns of a certain calibre necessary to do good service in the Field.—Permanent Field Artillery would rival the Horse Artillery in point of efficiency.—Want of time at present for such Instruction.—Proportion of Horses to the Men in the Artillery.—Necessity of separating Colonial from Field Artillery.—

Necessity of a Permanent Pontoon Train.—Experience necessary and principal objects to be accomplished by this particular Service.—Some observations on the Pontoon Service of the British Army in Spain and the Netherlands.

WITH respect to the want of uniformity in drills, duties, &c., briefly expressed in the pamphlet of 1849, the author had in view the difficulties caused by the detached system rather than any want of uniformity in the abstract; for instance, one troop (Duncan's) was formerly permitted to be essentially different from the others as to drills and exercises; and at this moment a different winter dress is allowed to be adopted for the Artillery in Canada from that in use in Nova Scotia. Nor was the complaint of want of unanimity intended to refer to anything like schism, but rather belonged to the same category as the preceding remark, for it is almost impossible that perfect unanimity can be maintained under the influence of the present isolated system; and if called upon to give an example, the author would refer to the conflicting opinions in the corps regarding the Horse Artillery as compared with the field batteries.

Want of uniformity and unanimity belongs to a company system.

Reverting to another question, esprit de corps, the writer considers its existence in the full meaning of the term to be out of the question whilst duty is done by companies; and considering its absence as a misfortune rather than a fault, he may venture to inquire what had

Supposed want of esprit de corps.

become of this feeling when eight field officers retired in 1843, and five others in 1850, when the duties in China and Jamaica were at the top of the rolster, neither of which circumstances could have occurred under a battalion system?

The last objection of any importance to which it may be necessary to reply, is the erroneous idea that the proposed change to battalions would cause a deficiency of officers as compared with the existing arrangement.

First, taking the colonial service—

	Officers.
The 48 companies proposed to be employed would, at their full strength, which they scarcely ever maintain, give	240
The field officers at the various stations are .	11
	255

The battalion system gives more officers at foreign stations, and

The 48 companies would, on the proposed system, become 72, or 12 battalions, with field officers: 12 × 24 = ...	288
According to the present arrangement . .	255
Gives . . .	33

more officers for the foreign stations.

So that there might be more than three on leave from each battalion, still retaining as many as belong to forty-eight companies of the present establishment.

But perhaps another comparison may make it still clearer that in forming smaller companies and troops, the working efficiency in the field is maintained or even increased.

Supposing a troop or a battery as in time of war to be increased by two officers, viz., a lieutenant-colonel and a lieutenant, it would then be equal to two smaller troops or batteries as the case might be, having, in fact, seven officers to the eight guns, instead of five officers to the six guns as formerly.

In the field also, than at present.

On a first view of one part of the question, the proposed field officers may appear to be too numerous; but the necessity for these officers for the artillery service is much greater in proportion to the men than in the case of cavalry or infantry. The guns and stores of forts and batteries involve a considerable and an admitted responsibility;* with reference to which, and to other circumstances, including the civil duties, one field officer to two companies (or, according to the present system, 200 men), is at any rate an established scale. Moreover, additional officers of this rank could with great advantage to the public service be employed at many stations at home; for example, in Alderney, Languard Fort, Weedon, Pembroke, and Chester, including Liverpool; also on the eastern coast, at several of the great estuaries which are neglected at present, and may be

Field officers required

at certain stations.

* Report of the Commissioners on Naval and Military Promotion, &c., 1840, p. x., pars. 5, 6, 7, 8, 9, 10, 11; also, pp. 55, 56, Nos. 917, 918, 919, 921, 923, 924, 925, 927, 928.

entered by a flotilla of armed steamers. Again, they are required at several places in Ireland, as Clonmel, Athlone, Waterford, Belfast, Cork Harbour, and possibly the important harbour of Bantry Bay. They could likewise be advantageously employed at several colonial stations: for instance, in New Brunswick, Newfoundland, and at the principal West India islands; also in Canada, the Ionian Islands, Malta, New Zealand, and Australia. They would also be of great use in assisting the commander of the battalion, whose duties as a respective officer often interfere with those which are purely military: a second field officer, as in the continental armies might, at the large stations, take charge of the stores under the Director-General of Artillery, with much advantage in point of economy.

The stores and equipments to be under the charge of field officers.

The preceding is merely an outline of the proposed change of organization in the artillery service; and it may be presumed that, on being reduced to practice, it would be found susceptible of much improvement.

The service should consist of horse artillery,

The branch of service under consideration should comprehend, as a portion of its force, a body of horse artillery, for the purpose of accompanying and supporting the movements of cavalry. A second portion should consist of light or field artillery, as the chief dependence for covering and supporting the movements of

light or field artillery,

the army; and a third, of the heavy artillery for garrison duties abroad and at home. As a part of the field service there should be a regular pontoon train.

garrison artillery, and a pontoon train.

At present we endeavour, and with a certain degree of success, to make every company fit for all kinds of duty, and the artilleryman becomes in consequence at once a cavalry and an infantry soldier; a grenadier at one time, by his size, and at another, from his activity, a light-infantry man.

Mixed duties of the artillery soldier.

On being dismissed from the marching and small-arm drill, he enters upon that of his own particular service, comprising the duties required both in the field and in garrison: stable duties are included in the former, and the use of all sorts of military machines constitute part of the latter. All are necessary for the complete instruction of the artillery soldier; but it is manifest that the efficient performance of such varied duties requires a separate distribution, or at least a subdivision into heavy and light artillery.

Separation into light and garrison artillery.

Men who are perfectly suited for the latter service, appear to require the youth, strength, and activity of light cavalry; and if such men were selected for field duties, they might, with advantage, as they became more advanced in life and less active in body, be transferred to the garrison artillery, in which they would, for several years, be fully equal to the duties, either

Gunners to be transferred to the garrison artillery.

at home, for example, at Portsmouth, Devonport, Guernsey, Jersey, &c., or at such places abroad, as Gibraltar, Malta, &c. As to the officers, the slowness of promotion has unfortunately left too many who are unfit for active duties; therefore, if enough can be found for the field batteries, there will be more than sufficient for the other branch.

Appointment of officers for the field service.

The officers required for a permanent field artillery, like the horse brigade, might be obtained by selection; the merits and qualifications of each individual being put in competition with that of others, on obtaining a higher grade. The opening to favouritism, which might be the consequence, has, with much reason, been objected to this arrangement; and this might easily be obviated by giving the vacancies in the field batteries to those who fall to them by promotion. But, on the other hand, it may be fairly presumed that the effects of an occasional instance of partiality would be more than counterbalanced by a general spirit of emulation and exertion to obtain the prizes.

Eighteen garrison and six field battalions of artillery to be formed.

Pursuing the idea of a separation of these duties, and recollecting that some of the foreign as well as the home stations require both field and garrison artillery, the twenty-four battalions might eventually be formed into six for the field, and eighteen for garrison service; which, with the present establishment of horses, would, as before observed, only cause an increase of 859*l*. 13*s*. 3*d*. in the annual expense, as compared with that of the officers and men already voted by Parliament.

Small expense attending the proposed change.

Three of the battalions to be formed at home might be immediately selected for field service—one and a half in Ireland, and the remainder in England; and in the course of two, or at the most three, years from this time, the whole regiment might be formed into corps of field and heavy artillery, and placed at the different stations at home and abroad, as in Schedule No. 6.

Recruiting to be general, to complete and fill up the regiment.

Recruiting, to fill up as well as complete, might be advantageously carried on for the regiment at large, by posting the recruits according to their fitness either for garrison or field service; including in the latter the horse brigade, the field batteries, and possibly a depôt of smaller-sized men, trained as drivers, to assist in completing the field batteries under pressing and particular circumstances. And, as it is proposed to transfer the men from field to garrison duties as they become inactive, the number specially enlisted for the latter service will, as a matter of course, be limited in proportion.

About 900 men,* or perhaps considering that

* The average number of deaths at present is 823.—Second Report on Army and Ordnance Expenditure, p. 56, No. 911.

many will not renew the period of service, 1,100, are annually required to replace the gunners who die or are discharged, and every effort should be made to obtain the same class as at one time flocked to the regiment. It is well known to old officers that there has been a great falling off in this respect since the peace, partly owing to the want of inducements, and partly to the terms of enlistment, as well as to the nature of the good conduct warrants. It is to be feared that the operation of the latter is but imperfectly understood, and that the account given of the protracted punishments, such as solitary confinement and hard labour, has an injurious effect. Limited service, with the prospect of being discharged after a period of twelve years, when the habits of the soldier unfit him, in a great degree, for resuming his former mode of life, appears to be another source of objection to recruits.

Objections to enlistment.

Since without due encouragement to the soldier that class of men cannot be obtained on whose conduct so much depends, everything should be done that may lead to the accomplishment of this desirable object; and if nothing else can be devised, offering equal inducements, occasional commissions to the deserving, would undoubtedly, as in former times, have a beneficial effect. For though, like the great prizes in other professions, they would be of rare occurrence, the bare

Necessity of encouragement.

possibility of advancement is absolutely necessary for the encouragement and well-being of mankind, and necessary above all to him who is ready to expose his life to danger in every form.

With reference to the preceding observations, the following remarks were made by Sir Hew Ross:*—

Remarks by Sir Hew Ross on this subject.

"Colonel Chesney has in his book stated, that he and other old officers look with deep regret on the present class of men joining the ranks of the artillery. That they are unlike those who formerly flocked to it, I will most boldly and decidedly declare from my own experience, having had charge of the recruiting service, that this is not the case. I could put a case to Colonel Chesney that would convince him that he is wrong as to what is the fact; and I appeal to any officer who looks at the men now assembled at Woolwich to say whether they are not equal, finer men, and better behaved than formerly."

The instance to which Sir Hew Ross probably alluded is freely admitted, as the exception, however, and not by any means as the rule in the present day.

Two circumstances have operated

In addition to the objections already mentioned as being supposed to operate with regard to protracted punishments, and the period of enlistment, there are two other causes which, in the opinion of the writer, have acted prejudically on the artillery service during the present as com-

* See Minutes of Evidence, Army and Ordnance Expenditure, 1849, p. 390, No. 6033.

against recruiting in the artillery.

pared with the last century. One is the altered prospects of the recruits, and the other the diminution of that class of the population from which they are intended to be chiefly taken.

Officers raised from the ranks in former times.

Concerning the first, it will probably be sufficient to mention the simple fact that at one time commissions were largely bestowed on deserving non-commissioned officers. The first company to which the writer happened to be posted was commanded by a deserving officer, Major Meredith, who had been raised from the ranks. And to show that this was by no means a solitary instance, it may be mentioned that, in 1795, the year of Sir Hew Ross's first commission, there were thirty officers of artillery alive who had been privates,* and subsequently to this period ten more were also promoted before the practice had ceased of giving away commissions to deserving men.

The manufacturing classes increased.

With respect to the second cause, it will probably be admitted by those who are best acquainted with the statistics of England, that the manufacturing has increased at the expense of the agricultural population. Previously to those changes which obliged the small farmers either to emigrate or to seek other employment at home, the labourers and farm servants belonged

* Kane's List of the Officers of the Royal Regiment of Artillery, pp. 3–51.

much more than they now do to the families of their employers. This particular class of persons is therefore greatly diminished, if not become almost extinct; and it is doubtful whether those who were not carried away by the manufacturing tide, can be as free from contamination as the farm labourers were before the change. But even if these circumstances have not caused any change in the character of the peasantry, the cessation of the ballot has operated, to a certain extent, against the artillery, since the fear of being drawn for the militia induced many to enter voluntarily, what was formerly a favourite service.

Soldiers recruited from the worst classes.

With regard to the general question of enlistment for the army, it has been stated by the highest military authority alive, that "the British soldiers are taken from the lowest orders of society. We only," says the Duke of Wellington, "get the worst characters, the most profligate and idle of their native village." *

The disadvantage is not so great in the artillery.

The writer, however, did not intend, nor was it by any means his desire to express himself so strongly with regard to the system of recruiting, for the artillery is still a popular service, though not so much so as when the boon of commissions and other inducements were held out; or when there was a better choice in the agricultural dis-

* Report on Military Punishments, &c., 1836, p. 324, No. 5805.

tricts. But when his statement, that there had been a falling off of late years in the artillery, was questioned, he forwarded to Lord Seymour and the Committee on Ordnance Expenditure a return showing the number of farm labourers and manufacturers composing the men of the two companies then under his command; and as these papers were not printed, as his explanation, in the Appendix of the Report, he avails himself of this opportunity to give an abstract of one of them as it stood at Cork harbour in May 1849.

Number of agricultural recruits,

Company.	Bakers.	Butchers.	Carpenters.	Smiths in general.	Weavers and Dyers.	Painters.	Tailors.	Shoemakers.	Masons.	Total Number of Trades.	Total Number of Labourers.	Number who can Read and Write.	Number who cannot Read and Write.	Read only.	General Total.
Capt. P. Ellis's .	2	.	5	2	6	2	3	3	2	25	30	33	19	3	55

as shown by a company of artillery.

As this company had recently returned from China, and was filled up after it came home, with the exception of the few gunners who had lived to return, it may be considered a fair average of the classes of men enlisted of late. Of the fifty-five, 25 were of various trades, with only 30 agriculturists of the number, and 19 could neither read nor write.

With reference to the state of promotion, about twenty-three field and other officers are employed at the Royal Military Academy, in the

Arsenal and elsewhere, including those staff appointments, which, like the adjutancies, take them away from ordinary duties for a considerable time. It is true that the outlet to be made by seconding those individuals would apparently be only temporary; but, as in the case of the engineer officers, it will probably be found that they will, generally speaking, be succeeded by other officers of the corps, and thus give to the latter the benefit of a certain number of steps, with, at the same time, the advantage of keeping the service complete in officers.

With reference to a proper complement of officers, it is presumed that the twenty-four officers allotted by the intended re-formation to a battalion would be sufficient as a war establishment, when it is augmented from 432 to 600 non-commissioned officers and gunners, which would give 14,487 non-commissioned officers and men, or, including the horse brigade, about 15,044 men; this establishment is about equal to the lowest estimate made by a talented artillery officer.*

The war establishment of twenty-four battalions = 14,487 officers and men.

It is, however, doubtful whether even this force would adequately provide for the wants of the service abroad and at home. Our numerous colonies require a serious augmentation of artillery, whilst the growing power of steam has

Deficiency of artillery in the British service.

* Sir Robert Gardiner proposed 10 troops of horse artillery, and 15 battalions of foot artillery.—Pamphlet of March 1848, p. 28.

made this arm a vital consideration for the defence of our coasts and harbours at home.

Necessity of increasing the British artillery.

Although it will be shown in these pages, and probably to the satisfaction of any dispassionate statesman, that a considerable increase of artillery is required for the service both at home and abroad, the writer does not propose to enter at much length into the extent of our wants in this respect. Leaving the details of that momentous question to be decided by those to whom it properly belongs, his observations will be chiefly confined to such a reconstruction of the regiment as would give to it younger officers, and greater efficiency in other respects.

A portion of horse artillery

With regard to one question. Strong and conflicting opinions are entertained by some of our ablest officers concerning the best description of field artillery, more particularly with respect to the ordinary batteries as compared with the horse brigade. But before entering into this discussion, the author would avail himself of the present occasion to rectify the misconception which seems to have existed in some quarters, that his opinion was adverse to the maintenance of the latter arm. How such an idea could have originated he cannot understand, since the necessity of maintaining this particular branch formed part of his former publication, and was pointedly repeated before the Committee of the

House of Commons. In No. 5854 he was asked,—

"Do you think the horse artillery essential?—I consider they are absolutely essential to the well-being of every army.

absolutely essential.

"5856. When you say they are essential for the well-being of an army, do you contemplate the use of six-pounders for the horse artillery, or do you contemplate their using larger guns?—The object of the horse artillery would be entirely defeated if they had larger guns. They once tried nine-pounders in Spain, but afterwards gave them up. To accompany cavalry, and perform the evolutions for which the horse artillery are really intended, the armament should be light.

"5876. You stated that you considered the horse artillery indispensable as an arm for the well-being of an army on active service?—Yes, absolutely necessary.

"5922. Do the Committee understand you to say that field artillery, according to your notion, may be made equally efficient with our horse artillery?—Not quite; the horse artillery have mounted detachments. In the batteries those men ride on waggons or walk; and for the most rapid movements they cannot equal the horse artillery: but for all the purposes of war, for the mass of artillery, they are perfectly equal to it, and at a cheaper cost.

Distinction between horse artillery and field batteries.

"5923. Have you ever known any officers of experience and authority who thought we could dispense with the horse artillery?—I believe there are several officers who think so; but I think the thing has not been understood, and I think that people have not considered what is the real object of the horse artillery, namely, rapid movements, getting on the flank of an enemy, and covering the passage of rivers, and other movements when we have not time to get up heavy guns. I think the services are distinct and very different."

A great, and it may be added an unquestionable authority in our service, thus grapples with the question of efficiency whether for horse or field artillery :—

"There can be no doubt that all guns of every calibre whatever, which the modern habits of warfare may bring into the field, are applicable to mounted artillery; the men of which service, though mounted for the sake of expeditious movement, are neither more nor less than other artillerymen the moment the guns are brought into action.

Two kinds of mounted artillery.

"Two ways of mounting the men attached to the guns seem to have presented themselves; one, by putting them on carriages, the other by mounting them on horseback. And according to one or other of these modes of organization, or to some modification of both, is the arm everywhere appointed. But in whatever way the men may be carried, in order that after a rapid movement they may begin unfatigued the laborious duty of working the guns in action, it is clear that the effect of the fire is the same. Once unlimbered, it is indifferent by what means the guns were brought, or the men carried to the assigned point; nor can there be any difference whether the men who work the guns belong to one branch of the artillery or to another. The only real difference will arise from skill, bravery, and previous instruction.

The men should be brought fresh into action.

Various expedients to lessen the expense.

"The additional expense attending the horse artillery beyond that common to other field guns, arising from the horses and appointments of the mounted men, has caused various expedients to be proposed for the conveyance of the detachments."*

It is scarcely necessary, with reference to the preceding, to observe that the "würst" on the

* Notes on Artillery, pp. 36, 37. London, T. Egerton, 1826. Understood to have been written by Sir Augustus Fraser, after his previous publication.

Continent, and with us, the four-wheeled waggon, answer the purpose in question, and it is well known that in our case, for the same calibre, the gun-carriages, &c., of the field-batteries are similar to those of the horse artillery; therefore the principal part of the equipment is a simple question, since the speed of a field-battery gun, that is, apart from its waggon, is, or *may become*, equal to one of the horse artillery.

But, as has just been seen, there is an important variation in the rest of the equipment, the gunners being carried on waggons with the former, and separately mounted to accompany the latter: this is attended with several advantages, and at the same time many serious disadvantages.

Perfection of the field batteries in Spain.

The advocates of the simpler equipment contend that, being capable of the same speed, the waggons might at all times accompany the guns, and that those brigades which had the requisite time for organization in the Peninsula, were considered, by the most competent authorities in our service, to be as perfect as the troops of horse artillery.

On the other hand, it is contended that the weight of the waggons with the men is fatal to speed; and that a battery, after the usual short period of instruction, can no more be compared with a troop of horse artillery, which is considered the very life-blood of our service, than a regiment of cavalry recently mounted can be with one which has perfected its discipline.

Superiority of the British horse artillery equipments.

Having had an opportunity of seeing the artillery of some of the continental services, and also that which has been so highly distinguished of late in the East Indies, the writer ventures to observe, that in point of efficiency a decided preference must be given to the equipment of the British horse artillery, not only over our own field batteries, but also over the élite of the several continental nations: and he offers this observation, notwithstanding a remark made to him in 1839 on the plains of Adrianople, by an officer of rank, regarding the corresponding portion of the Russian army then manœuvring there, "Que c'était tout qu'il y a de plus parfait dans le monde."

Conflicting opinions about the best kind of field equipment.

The question so long at issue is not, however, the incontestable efficiency of the Royal Horse Artillery, but whether, as has been alleged, permanent field batteries would, by answering every purpose, enable the Government to dispense with the more expensive equipment; an opinion which, in the judgment of the writer, is open to the same objection as that of those *who would only have horse artillery*; misconceiving the object to be attained by that child of Frederic the Great, which at a later period was so carefully fostered by another great captain in the art of war—Napoleon.

Field batteries in France in 1818, and

The efficiency of the British field batteries attracted the particular attention of the French officers of artillery in 1818, at the grand review which preceded the departure of the army of occupation from France.

"When the cavalry were drawn up in two lines on the plain between La Selle and L'Ecaillon, and supported by masses of infantry, upwards of a hundred pieces of horse and foot artillery were seen to issue at the same moment from the different bodies of troops, and move forward *with equal velocity* a hundred toises in front of the first line of cavalry.

"The affair commenced by a cannonade resembling a running fire, and the movements of the artillery were executed with great order and precision. The facility with which the guns were limbered and unlimbered did not escape the notice of our artillery officers.

their great efficiency.

"One of the brigades of foot artillery met with an accident in passing, at full gallop, a very precipitous road. Having found the passage obstructed by a calèche, the brigade dashed rapidly into a field on the left to avoid any delay: only one of the guns was, however, overturned, and but one of the gunners suffered from its fall; the piece was righted in an instant and joined the brigade.

Movement of the cavalry.

"The two lines of cavalry charged twice in succession, and, as if repulsed, rallied alternately behind each other. The gunners then, retiring rapidly with the limbers, deserted the guns, and presently afterwards returning to the charge recommenced firing. The passage of the Ecaillon was supposed to be defended by the enemy's army occupying the villages, bridges, and all the important points of the position, suggested the plans and evolutions necessary for an operation of this importance. The passage being forced, the supposed enemy, finding his left flank turned, and being pressed by the allies, commenced a retreat, contesting every foot of ground, and defending every village, until he reached the camp on the heights of Famars. The artillery followed the troops through the steep valley of L'Ecaillon, and through the plain beyond (overcoming every obstacle), to the foot of the heights of Famars, arriving on the ground at the same moment as the rest of the army. This position is covered by a deep ravine difficult of ascent, ***which the artillery cleared like the cavalry.*** Some carriages were overturned, but quickly put upon their wheels again. Our artillery would not have surmounted this obstacle without much time and labour." *

Passage of the Ecaillon.

Guns overturned and speedily righted.

At the period in question there had been sufficient time for the officers and men attached to the field batteries to become acquainted with the various field duties, including that part of them on which so much depends—the care and management of the horses. Doubtless, *permanent* field batteries would attain an equal, if not a greater degree of efficiency; and bearing in mind that the officers and men are from the same stock as those of the horse artillery, batteries, when of the same calibre, would, with the single difference between carriages and mounted detachments, rival the troops of horse artillery, and have at the same time those advantages over this arm which are claimed by their advocates; such as taking a larger supply of ammunition with fewer horses and men exposed in action, and, what must be an important consideration, with less demands upon the Commissariat during a campaign.

Permanent field batteries.

Field batteries possess certain advantages.

* Force Militaire de la Grande Bretagne, par M. Dupin, pp. 229, 230.

With reference to these advantages, it may be mentioned that, on carefully weighing the field equipment at Ballincollig, when complete in every respect for service, it was found that the 6-pounder, with two gunners in marching order on the limber, weighed 32 cwt. 3 qrs. 25 lbs., and the waggon, carrying six men, 38 cwt. 1 qr. 8 lbs.; thus showing that the latter is 5 cwt. 1 qr. 11 lbs. heavier than the former.

Weight of guns and carriages.

Now, if it were to be considered an object of paramount importance to equalize the weight of the gun and its waggon, either by constructing a lighter carriage, or by leaving a small portion of the ammunition with the reserve,* it is obvious that the speed of the latter could in every case not only equal that of the former, but also that of the horse artillery gun itself, from which, in point of fact, the one used in the field batteries does not differ. Bearing this in mind, the advocates of the field batteries are justified in contending that, when of the same calibre and similarly horsed, this branch is quite capable of maintaining the same speed as the horse artillery; and if it be admitted that, with the same degree of pains and care, similar guns and carriages will have equal speed, it will scarcely be denied that when reduced to the same weight the waggons will move

Field battery guns, having waggons of the same weight,

* Mounting gunners on the off-horses, as in the Bengal Horse Artillery, offers another available resource.

as rapidly as the guns to which they belong. It is true that carriages without springs are a severe trial to the men; but it may be observed that the two horse artillery gunners on the limber are exposed to the same inconvenience.

But without attaching too much importance to what has been considered a proof that equal efficiency with the horse artillery *may* be attained by the field batteries at much less expense, it cannot, and probably will not, be denied that an efficient battery, by the addition of mounted detachments, might at any time be converted into a troop of horse artillery; at the same time it must not be forgotten that much depends upon this distinctive difference between the two equipments. The extra horses may be applied in draught when necessary, and thus facilitate the passage of difficult ground with the speed requisite for the accomplishment of the principal objects of flying artillery, viz., to be ready to advance rapidly when and where such guns may be employed to gain the flank of the enemy, to seize a particular post, such as a defile, the passage of a river, or to cover advanced movements; also to accompany a flying expedition with light cavalry, to reinforce any part of the line that may be threatened, and, finally, to complete a victory by the rapid pursuit of a retreating army.

would equal the speed of the horse artillery.

Particular objects to be accomplished.

For these important objects horse artillery is just as important to the well-being of the artillery service as the cavalry is to that of the army itself; but recent experience has shown that the requisite speed *cannot be maintained* with heavier guns than 6-pounders. This is also the opinion of a distinguished officer, Sir Robert Gardiner, who has set this question at rest by saying "that the necessary quick movements of the horse artillery could not be attained by 9-pounders; the telling effect of 9-pounders could not be expected from horse artillery."* The want of speed of the larger calibre has recently been shown in the field. A troop of horse artillery, armed with 9-pounders, attempted to accompany the cavalry sent, after the battle of Goojerat, in pursuit of the enemy, with General Gilbert's force. They commenced at a gallop, but came down to a canter, a trot, and at last to a walk; the horse artillery, armed with 6-pounders, passed them at a gallop, and only pulled up when the cavalry pulled up, and was with them at the very end of the pursuit.†

Importance of horse artillery to an army.

Light guns are indispensable to speed,

It is true that field batteries, with guns of the same calibre and lightened waggons, would have done the same, and also that they would be much

though great speed may be attained by light field batteries.

* Report on the Numerical Deficiency, Want of Instruction, and Inefficient Equipment of the Artillery of the British Army, &c., by Major-General Sir Robert Gardiner, K.C.B., R.A., 1848, p. 34.

† Major-General Parker's Evidence, Second Report on Army and Ordnance Expenditure, July, 1849, p. 454, No. 9010.

more economical during peace, and easily converted into troops of horse artillery on the approach of hostilities. But as no army can have the requisite efficiency without cavalry and a fair proportion of horse artillery, for the objects already mentioned, for which the extra horses, whether in draught or as reserve, give an immense advantage, it would be ill-timed economy to dispense altogether with the latter in time of peace, when the appointments to this arm are at once rewards for past services and the means of stimulating officers to similar exertions in future. But while retaining a due proportion of horse artillery, it is of still more consequence to secure the efficiency of the field batteries, on which, much more than on the former, the result of pitched battles largely depends.

Horse artillery should be maintained,

and field batteries made efficient.

Evidence of certain officers proposed to the Committee.

Perceiving from the evidence of Sir Hew Ross, Lieutenant-Colonel Brereton, and Major-General Parker, that their attention was chiefly confined to the horse brigade, the writer ventured, in a note, to suggest to the committee, with reference to the latter service, the names of Major-General Paterson; Lieutenant-Colonel H. G. Jackson, since deceased, who commanded the artillery at Crystler's Farm, in America; Lieutenant-Colonel R. G. B. Wilson, who served in a battery at Waterloo; and Captain Piercy Benn, then in command of a battery at Charlemont, and previously one of the adjutants at Woolwich. Had these officers been called, it i probable that they would have shown, as well as felt, that the field artillery of Great Britain requires something more than some troops of horse artillery, however perfect in themselves. For, it may be observed, our horse artillery is, but our field batteries are not, what they are capable of, and ought to be made.

The necessity of horse artillery being admitted, and it is hoped established, the proportion that should be maintained alone remains to be considered.

Proportion of horse artillery for an army.

In 1833, there were allotted to a Prussian corps d'armée, consisting of 28,000 men, three troops of horse artillery and nine batteries.* It is, however, understood that the number of the former, as compared with the latter, has been diminished of late; whilst one-fifth is the full proportion in France. This would give about seven troops of horse artillery for an army of 60,000 or 70,000 men, or nearly that which has been thought a fair proportion for any British force that is likely to take the field.

Proportion required for 60,000 or 70,000 men.

Armament of the field batteries.

Reverting to the armament of the field batteries, 9-pounders having been used so successfully at Waterloo, and again adopted during the later service in Portugal, they will, as a matter

* Prussia in 1833, by M. de Chambrey.

of course, be continued: but it should not be forgotten that 12-pounders form a considerable portion of the continental field artillery, and that much of our loss on the 18th of June, 1815, was caused by the fire from guns of this calibre; these may, therefore, be partially introduced with great advantage, and a larger proportion of howitzer guns and rockets seem to be advisable also as part of the field service. Batteries of 12-pounder howitzers, for instance, are about to be introduced into the French service; and rockets, which for some time have attracted much attention, particularly in the Austrian service, are doubtless destined, on Hale's principle, to take a prominent place in the next war.

Howitzer guns and rockets to be increased.

The late Sir Augustus Fraser, in his second pamphlet, entitled Notes on Artillery, page 36, observes,—" It seems peculiarly necessary that though some portions of mounted artillery, in reference to the duties of light cavalry, may be armed with light ordnance, the majority should be equipped with powerful and efficient calibres, such as may do great and powerful execution." But if this seemed to be necessary formerly, it becomes of the last moment now, as the means of effectually opposing the new musket which is about to be adopted in European armies.

Field guns to be chiefly of heavy calibre.

In making the field artillery permanent, as above proposed, and at the same time a separate service, as much so, at least, as the horse artillery is at present, there can be but little doubt that the efficiency of the latter would be closely approached by the field batteries: at present, this is out of the question, owing to the want of the necessary time for instruction, and the constant changes of officers and men which become unavoidable in attempting to make the whole of the ninety-six companies equally fit for field and garrison duties; or, in other words, to make all the men at the same time both cavalry and infantry soldiers; while, on an average, there is scarcely at present more than one horse for about twelve men in our service.*

Separation of the field and garrison service desirable for the efficiency of both.

The duties of the colonial and field artillery are so different, that it is an object of paramount importance for the well-being of the service that the two branches should be separately organized: the field artillery should comprise horse artillery as well as field batteries; and, as part of both, there should be a regular pontoon train.

The field service to be officered by selection, and

The officers for the field batteries, like the horse brigade, might be completed by selection, their merits and qualifications being put in competition with those of others, on aspiring to higher grades. It would, however, be desirable, as a

* There are, exclusive of the colonial artillery, 10,603 non-commissioned officers and men in the regiment, and 928 horses in the several field batteries at home and abroad. Ordnance Estimates, 1850-51, pp. 6, 10.

officers who are employed to be seconded.

matter of justice to the service, to give temporary promotion throughout the regiment by *seconding* those who are likely to be absent from duty for a certain length of time, from any cause whatever.

The materials of our arsenals are ample and well suited for this purpose, but more practice is desirable; and it is deserving of serious consideration, whether there should not be a portion of the service specially allotted, and constantly trained to the management of every kind of bridge equipment, as has long been the case with other nations.

A pontoon train.

In France the personnel and matériel for the construction of military bridges are extensive, and are fostered with extreme care, which, however great it may appear, is not more than is absolutely required to accomplish the objects on which the success and even the safety of armies so frequently depend. This being the case, the pontoons have belonged to the artillery service since the beginning of the fifteenth century, and the permanent bridges to the engineers.

It may not be altogether out of place here to observe, that during the service intrusted to me by His late Majesty, a variety of circumstances occurred which showed the necessity of such an organization.

The materials for the construction and armament of two steam vessels, amounting to some hundred tons weight, were to be transported a distance of 145 miles across a country presenting various features. Some parts were hilly, others level and marshy; there was also an extensive portion of lake navigation, and there were two rapid rivers to be passed. The latter were, at all times, nearly impracticable, being occasionally so deep that some of the carriages were actually dragged from bank to bank under water; yet the heaviest weights were safely transported from the Mediterranean Sea to the Euphrates. Various means, according to circumstances, were used, such as Blanchard's pontoons, flying bridges, launches, and lighter boats, some of which were of canvas; gins, hawsers, jack-screws, and finally animal power, were also brought into requisition. In one place, of exceeding difficulty, the boiler of the "Tigris," rather more than seven tons weight was removed by the late Lieutenant Cleaveland, R.N., in an extraordinary manner. The bullocks of this part of Syria being small, and but little accustomed to draft, no fewer than 104 of them were yoked four abreast to the truck, and they were conducted by fifty-two people of the country.

Instances given to show its importance.

Means of crossing rivers.

Thus terminated eight months of unceasing exertions, during which the officers and men of the expedition encountered every difficulty (open hostility alone excepted) which belongs to the transport of the equipments of an army. But with the exception of the substantial floating bridges, which have probably been used in Mesopotamia since the time of Xerxes, all the measures adopted were new to the party in Syria, or at least they had not been practically tried.

Establishment of a pontoon train in the service.

It is, therefore, hoped that the experience then gained will justify the recommendation which the writer has ventured to make, that an efficient pontoon train should form part of any new organization of the artillery service that may seem advisable. One of the proposed battalions would probably be sufficient for this purpose, and it might either be attached to, or form a part of, another important branch of the service, namely, the field train.

Principal objects to be accomplished by a well-equipped pontoon train.

The objects to be accomplished by the establishment of a pontoon train, relate, 1st, to the positions occupied by, and the principal communications of, an army, and comprehend the formation of the more substantial bridges, as those on piles and tressels, or those of boats, pontoons, casks, &c.; 2ndly, to those connected with its movements, more particularly with the operations of cavalry and light troops, and comprise the formation of frame or lever bridges suited to pass narrow rivers, bridges of cordage,

&c., stretched from side to side for wider streams; also, for the same purpose, flying bridges of rafts and small pontoons, inflated skins, or prepared canvas cloth. These means, or a part of them, should be at hand with the advance, so as to be always available.

Advantages of previous instruction in pontoon service.

Gratefully recollecting the advantages derived from the practical instruction received at Woolwich and Chatham by the detachments of artillery and sappers attached to the Euphrates Expedition—which instruction became, with a little subsequent organization, (thanks to the unceasing exertions of the officers of the expedition,) the means of accomplishing the transport of the stores, one of the most gigantic operations of modern times—the writer has endeavoured in the preceding remarks to show the desirability of having a pontoon train in the British service, permanently equipped for drill and instruction. This might be either attached to the engineers, or, as in the French service, form an integral part of the artillery, which, as the most cumbrous portion, requires a pontoon train more than any other part of the service. It would in this case, as an indispensable adjunct, be always ready for any future campaign.

In France, the pontoon train is attached to the artillery.

The want of practical knowledge in the management of such equipments is dwelt upon by Sir Robert Gardiner in his pamphlet; and the letters published in the "Morning Chronicle" by Major-General Sir Charles Pasley, who had much to do with the pontoon service, show its defective state in the earlier part of the Peninsular campaigns, and also that much still remained to be done at their termination.

State of the British pontoon service in Spain.

"Early in 1812," says Sir Charles, "Lieutenant Piper, of the Royal Engineers, was placed in charge of a train of pontoons recently sent from England, which was drawn by oxen supplied by the Commissariat. It was guarded by Portuguese infantry, and the only pontooneers were a few English artificers to keep the pontoons in repair, and a party of Portuguese seamen. With these means, however, a pontoon bridge was placed over the Guadiano, by which 12,000 men passed the river on the 16th March, 1812, to besiege Badajoz." *

"Early in 1813, larger pontoons from England were substituted for the smaller ones first used, and formed into two divisions of 18 pontoons each, one under the same officer, and the other under Captain English, of the Royal Engineers. The Portuguese seamen were now increased to 100 men, under a lieutenant and two midshipmen from that service; and horses, under Lieutenants Wilford and Matthison, of the Royal Artillery drivers, were substituted for the oxen.

Passage of the Garonne delayed in consequence.

"The Duke of Wellington, being obliged to postpone his attack on the French army in consequence of the delay that occurred in throwing a bridge over the Garonne in April, 1814, and perceiving that the state of the horses had much to do with the impediments, Captain Green, of the Royal Artillery, was ordered to take charge of the drivers and horses,

* Major-General Sir C. Pasley's Letter, published in the Morning Chronicle of the 5th July, 1849.

the scientific duty being still intrusted to the engineer officers."*

The pontoon organization incomplete in 1815.

But as hostilities terminated within a fortnight of the change here alluded to, it is obvious that the pontoon train was only in process of organization; and it is equally clear that up to 1815 this desirable object had not been accomplished; for the Duke of Wellington appears first to have contemplated the employment of a distinguished post-captain of the navy (Sir Charles Napier) and 200 seamen; but on finding that some of the Sappers and Miners had already been trained to this duty,† their services were declined, and the whole department was then organized on an extensive scale in the short space of two months; but still, as it appears, with the disadvantage of having Flemish drivers.

* Major-General Sir C. Pasley's Letter, in the Morning Chronicle of the 23rd June, 1849.

† Letters of the Duke of Wellington to Earl Bathurst, dated 2nd and 12th of May, 1815. Gurwood's Despatches.

CHAPTER VII.

ON INCORPORATING THE ORDNANCE CORPS WITH THE ARMY.

Objects of the proposed Reconstruction.—Artillery hitherto scarcely an accessory of a British army.—Employment and uses of Reserve Artillery.—Opinion of Sir Augustus Fraser on the employment of Field Officers of Artillery.—Proposed Centralization of the Artillery.—The Military Officers of the Ordnance should be eligible for General Employment.—Evils of the present System of Exclusion.—If placed under the Commander-in-Chief, many disadvantages would be remedied.—Earl Grey's opinion about a change of system.—Mr. Hume's Memorandum on the Ordnance.—Field Train Establishment proposed by Sir Augustus Fraser.—Officers employed in the Field Train should be Military rather than Civilians.—Outlet for deserving Non-commissioned Officers.—Opinion of the Committee of the House of Commons on this subject.—Mr. Hume's ideas, and the proposed Sections of Artillery under Lieut.-Colonels.—The Duke of Wellington complained of the Deficiency of the Artillery in Spain.—Board of Respective Officers at Hong-Kong and their Duties.—Orders issued by Major-General D'Aguilar, suspending two Members of this Board.—Appeal to the Home Authorities.—Attack on the Author and others by the Chinese of Fuchan.—Expedition to the Canton River.—Reports of the Respective Officers.—Result of the Appeal to the Home Authorities.—Proposed change regarding the Duties of the Respective Officers at Foreign Stations.—Necessity of Younger Field Officers.—All Military operations should be under one head.—The strength of the Artillery should be regulated by that of the Army.—Observation of the Committee of the House of Commons.—General Centralization.

Reorganization of the artillery service.

But to return from the subject of a pontoon train to the main question of reorganization. The change proposed in the preceding chapter has chiefly two objects in view.

1st. An efficient field artillery; and 2ndly, as

indispensable to this object, and to the general service of the artillery, superior officers who are still in the prime of life.

Objects to be accomplished.

With regard to the former, one step, and perhaps it may be said only one, has been made from the inefficient fire of battalion guns, to the employment of batteries with certain portions of the army in the field. However efficient these isolated sections may have been, it is no disparagement to zealous officers to add, that the artillery in Spain was merely an accessory, and not an arm, being entirely deficient in that higher organization, on which decided and brilliant results so much depend. But let the great experience and high authority of a distinguished officer illustrate this point:—

"Let us consider," observes the late Sir Augustus Fraser, "what is the present situation of the officer commanding the artillery of any army.

Anomalous position of field officers of

"He is expected to be responsible for all that is understood by the efficiency of the arm; yet he has nothing to do with its instruction before it joined the army, when it is subdivided and placed in various ways under the command and superintendence of the general officers of the cavalry and infantry: so that, with the exception of attending to its wants in men, horses, ordnance, and stores, and of endeavouring, by correspondence with the department in England, to obtain the necessary supplies, the commanding officer of artillery may be almost said to have little to do with the arm in the field. His opinion, we have seen, is not asked as to the selection of the arm for service, and it is as clearly never required for the distribution of it afterwards. He can seldom be known to his commander but by the wants of the arm, or by the sins of the system. Can it be wondered then, that, rarely coming in contact but on these ungracious occasions, he seldom obtains the consideration which he does not appear to deserve? As to signalizing himself by any happy application of the arm in the field it is out of the question. He remains an individual, without the power of moving a single gun.

artillery in the field.

"With respect to the field officers of artillery with an army they are, by the present distribution of the arm, placed in a very singlar situation. They are appointed to, and understood to have, the command of two brigades of field artillery. But these brigades, being attached to a division of the army, are habitually separated from each other. In consequence, the field officer cannot be with both: his presence with either is displeasing to the officer commanding the brigade, who naturally wishes to receive the credit of acting independently; and the field officer is reduced to the alternative of either doing nothing, or of interfering with the command of a captain, probably very competent to the charge of his own brigade. Even in the ordinary routine of transmitting the daily states and reports of the brigades, the habit of the present day is for the captains of brigades to send them direct to head-quarters without their even passing through the field officer, who is, and who feels he is, a mere cipher.

"Now, if the service were benefited, or the field artillery made more efficient by this virtual suppression of the field officers, there can be no question that these officers should not stand in the way; there can be no doubt that their interests should yield to the paramount interests of the service. But if, so far from benefiting the service, this supercession be evidently one of the very causes of the want of efficiency, and if the instruction of which the corps stands in need ought to spring from these very officers, this ruinous system, which is fatal to anything like zeal, cannot be too soon done away with.

Want of supervision.

"It requires, indeed, no argument to prove that, in order to have any organized body, there must be a gradation of ranks, and that these can never be inverted without doing real injury."*

Necessity for a gradation of ranks.

Instead of continuing the preceding system, by which a field officer *nominally* commands two batteries serving in different divisions or brigades of the army, without belonging to either, one of centralization has been taken as the basis of the present organization. A battalion of artillery would therefore take the field *as a whole*, and the detachment of batteries become the exception, leaving at all times a powerful reserve.

Troops acting in masses, accompanied by artillery.

Agreeably to the tactics of the present day, of troops acting in separate masses, the cavalry and infantry must have a portion of artillery attached to each, in order, amongst other things, to accomplish the important object of dislodging an enemy from buildings or other obstacles, which would be almost irresistible to cavalry or infantry alone. When it is necessary to strengthen the reserve artillery to carry or defend some vital point, these portions would necessarily be concentrated; but under ordinary circumstances, the reserve artillery is supposed either to act by itself, or to support either of the other two arms as the case may be. The movement of this, and the other portions of artillery in battle, should

Employment of reserve artillery.

* Remarks on the Organization of the Corps of Artillery in the British Service. Rowland Hunter, London, 1818, pp. 65–68.

therefore be rapid, and have a definite object as part of the general's plan. It is consequently indispensable to the success of the latter, that the commanding officer of artillery should be made aware of, and should also be capable of entering into the various combinations that may be intended, whether for attack or defence. A well-organized artillery should therefore be ready to act as a separate arm, and should not only take its share during the heat of the battle, but be capable also of pursuing a retreating enemy in order to complete the victory. Since, in the judicious employment of this arm, much depends upon its being brought into action unexpectedly, speed in taking up a position, whether by horse or field batteries, is indispensable. The gunners, therefore, of the latter, although marching at other times, should be carried on their waggons at such a moment, in order that they may reach the intended position not only quickly, but sufficiently fresh to do their duty with effect. Being capable of very rapid movements, the batteries in Spain and at Waterloo were to all intents and purposes second-rate horse artillery; and being so, it need scarcely be observed, that if the gunners were to be on foot, when moving in battle, the guns must impede the movements of the army.

Combined services of the artillery.

Gunners should be carried into action.

With regard to the other main object, the physical efficiency of officers, it is hoped that, after a little time, by means of some such organization as that shown in Schedules 2 and 5, officers would obtain the rank of Colonel when about fifty, and that of Lieut.-Colonel between theages of thirty-five and forty. It has been well said that no man will cheerfully do his duty, unless he has the prospect of future reward for his services. As an additional and powerful incentive, therefore, to the efficient exertions which may be expected where youth and science go hand in hand, the artillery and engineer officers should be permitted to serve on the general staff, sharing at the same time other duties, at home and abroad, equally with the officers of the line. The exclusive nature of separate services has been attended with many disadvantages; but these it is hoped, will soon be remedied, by their becoming one service under the Commander-in-Chief.

Staff and other duties to be open to the artillery service.

The artillery and engineers to form an integral part of the army.

As the questions put by the Committee of the House of Commons did not afford an opportunity of entering into this subject, it may now be proper, although necessarily at some length, to endeavour to show some of the disadvantages caused by maintaining the military branches of the Ordnance as separate services.

Disadvantages of separate services,

Although there are many advocates of the existing system among the senior officers of both corps, it would assuredly be considered an anomaly in any other army, if the artillery and engineers, when out of the kingdom, were entirely under the officer in command of the troops, whilst at home they are independent of the Commander-in-Chief, with, at the same time, all the disadvantages attending additional references concerning leave of absence, changes of station, reliefs, and every kind of duty.

But the counteractions to which the workings of this system are liable, will be most forcibly shown by a paragraph quoted by Earl Grey on the centralization of departments:—

and separate boards, as shown by Earl Grey.

"It is, however, in our opinion, a much stronger objection to the existing system than that to which we have just adverted, that it causes an inconvenient separation, not merely of accounts, but of the management of different branches of the same service. It is impossible not to remark, that various duties, which all have reference to one common object, and in the discharge of which it is highly important that there should exist the most complete unity of purpose, are intrusted to authorities not merely separate and distinct from each other, but mutually independent, and only connected together by their common subordination to the supreme authority of the Government. Under this system, however anxious those who conduct the several departments may be to keep up a good intelligence with each other, we believe it to be impossible that a want of due concert and vigour in their various measures should fail to exist; and, accordingly, we are much deceived if the practical results of the absence of a more concentrated authority are not to be traced in conflicts of opinion, diversities of system, and delays exceedingly injurious to the public service, while we think there has also been some un-

Want of concert and vigour of action.

necessary expense of establishment, and a good deal of multiplication of correspondence and of needless formalities in the transaction of business."*.

Multiplication of correspondence and other evils.

Evils such as these will probably find a remedy at no distant day, by the appointment of a War Minister to take charge of all the departments, the Ordnance included. With respect to the Ordnance corps, the Report from the Select Committee on Army and Ordnance Department, 1849, Appendix K, p. 1079, on the same subject, contains the following:—

Remedies proposed by Mr. Hume.

"MEMORANDUM respecting the Ordnance, for consideration.
(Proposed by Mr. Hume.)

"1. To transfer the command and discipline of the Royal Regiment of Artillery, and Royal Engineers and Sappers and Miners, to the Commander-in-Chief.

"2. To appoint a Deputy Adjutant-General from the officers of the Artillery to the staff of the Commander-in-Chief, to aid in performing the Artillery staff duties, but to be considered as one of the staff of the army, available for carrying out the orders of the Commander-in-Chief, whose orders may also be conveyed to the Artillery through the Adjutant-General, as the staff at head-quarters.

Proposed staff of the Ordnance.

"3. To appoint an officer of the Royal Engineers to be an Assistant Adjutant-General at the Horse Guards for the duties of the Royal Engineers and Royal Sappers and Miners, in like manner as for the Artillery.

"4. Both these officers for the Royal Artillery and Royal Engineers to be considered entitled to advancement on the staff.

---

* Minutes of Evidence taken before the Select Committee, &c., July 5, 1850, p. 700, No. 9040.

Second captains of artillery and engineers to be abolished.

" 5. To appoint a Commandant of Artillery and a Commandant of Royal Sappers and Miners, each with a staff of an Assistant Adjutant-General and Brigade-Major respectively, for the transaction of all regimental duties.

" 6. To render officers of Royal Artillery and Royal Engineers available for all staff employments, in like manner as obtains for officers of the army.

" 7. To reduce all Second Captains in the Royal Artillery and Royal Engineers; but to render the Royal Artillery and Royal Engineers at all times perfectly efficient in officers of the rank of Captain; strictly, to fill up the vacancies of all officers of this rank absent on any duties, other than regimental, by being seconded whilst so absent, and when they return to their corps the next vacancies occurring in their rank not to be filled up.

To do duty by troops and companies.

" 8. To abolish all battalions in the Royal Artillery, and to do the duty by troops and companies, the organization of which to be made complete and efficient in all ranks: efficient depôts of instruction for riding, for drilling, for forming recruits, and for receiving men absent from their companies, allowing a staff for regimental detached commands.

Colonels-commandant to be discontinued.

" 9. To strike off from the Ordnance corps all Colonels-Commandant, and to reduce the Colonels and Lieutenant-Colonels to the same proportions as the ranks of Colonels and Field Officers bear in the other branches of the army service to the other ranks.

Full-pay retirement of officers to be permitted.

" 10. To form a retired list from the Royal Artillery and Engineers, granting to all officers increased retiring pensions, according to length of service, or according to rank, as may be most favourable to the officers, in proportion to the advantages which Colonels of regiments of cavalry and of the line at present enjoy.

Garrison artillery to be formed of worn-out men, &c.

" 11. To form a garrison artillery of officers and men who may be unfit for all the active and laborious duties of the field artillery.

"12. To transfer the horse artillery to the field batteries, making these so efficient as to ensure at all times . . . field batteries, each with . . . * horses, being kept up in the United Kingdom, and so constituted as to be capable of rapid increase.

"13. To arrange for placing ordnance and stores as much as possible under regimental charge of regimental officers, affording the means of performing the duties connected therewith. Ordnance stores to be under regimental charge.

"1st. To reform the Ordnance Department by constituting it of the following ranks:—

1. Storekeeper-General, or Principal Commissary of Ordnance, having a general control over all stores and depôts.
2. Storekeepers, or Commissaries of Ordnance. Proposed field train.
3. Deputy Storekeepers, or Deputy Commissaries.
4. Assistant Storekeepers, or Assistant Commissaries.
5. Conductors of Ordnance.
6. Sub-Conductors of Ordnance.
7. Store Sergeants.
8. Store Corporals.
9. Store Men.

"2nd. The office of Storekeeper-General, or Principal Commissary of Ordnance, to be filled by an officer of Artillery chosen from any rank. The storekeeper-general, commissary, and other

"3rd. The situations of Commissaries and Deputy Commissaries to be open to Captains and subalterns of the Artillery; and one-half at least of these situations to be filled from the corps of Artillery, but not to be held longer than for a period of five years, when they will revert to their regimental duties.

"4th. In order to improve the condition of the soldier, all

* These numbers were apparently left by Mr. Hume to be filled up by the authorities.

such appointments to be given to artillery officers.

the other situations to be filled up from the ranks of the army, giving to the Artillery and Sappers two-thirds, and to the other branches of the service one-third, of all the appointments."

Duties of the Chief-Commissary,

The higher duty of the Chief Commissary and Commissaries of the Ordnance Department need no detail, since they are well known to embrace all that is connected with the supply and expenditure of Ordnance and Artillery stores of every description. Much therefore obviously depends on the ability, zeal, and integrity of these officers.

The Assistant-Commissaries are placed in charge of depôts and reserve ammunition. They have the equipment of battery trains and of field ordnance; and are, in many cases, the officers to whose exertions the commanding officer of artillery can alone look for the supply of those ordnance equipments, on the prompt arrival, and complete state of which, military operations of considerable importance may depend.

The duties of conductors of stores are principally those of loading and unloading vessels, escorting stores on a march, and superintending artillery working parties.*

and officers of the field train.

Such being the employment of the field train, it seems deserving of consideration whether this branch of the service might not give a more

* Remarks on the Organization of the Corps of Artillery in the British Service. Rowland Hunter, London, 1818, pp. 129, 130.

extensive outlet than that suggested by Mr. Hume, and renew in another shape those hopes which belonged to the artillery soldier in former time. Nor can it be questioned that a service of this kind, composed exclusively of military men of suitable habits and acquirements, would by their practical knowledge of the artillery service have a decided advantage over one composed, as at present, of civilians.

Past and proposed establishments.

Alluding to the expense entailed by the number of officers required in consequence of the inexperience of the department at that time,* it has been proposed that, instead of the war establishment of 1812, which was—

		Per diem.	£.	s.	d.
1	Chief Commissary,	at 30*s*. . .	547	10	0
3	Commissaries,	at 20*s*. . .	1,095	0	0
3	Commissaries,	at 15*s*. . .	821	5	0
6	Assistant Commissaries,	at 10*s*. . .	1,095	0	0
6	Ditto ditto, 2nd class,	at 8*s*. . .	876	0	0
50	Clerks of Stores,	25 at 7*s*. . .	3,193	15	0
		25 at 6*s*. . .	2,737	10	0
60	Conductors of Stores,	30 at 5*s*. . .	2,737	10	0
		30 at 4*s*. . .	2,190	0	0
129			£15,293	10	0

* It is, however, well known that the number of the officers of this department, and in consequence of the expense of it, have much exceeded what was anticipated; nor does it appear unfair to imagine that the increase of officers has, in many cases, been occasioned by the inexperience or incompetence of several of them to execute duties requiring

The field train in war time might be—

Proposed employment of artillery officers,

		Per diem.	£.	s.	d.
1	Chief Commissary,	at 50s. . .	912	10	0
3	Commissaries,	at 30s. . .	1,641	10	0
3	Deputy Commissaries,	at 25s. . .	1,368	15	0
12	Assistant Commissaries,	at 15s. . .	3,285	0	0
36	Clerks of Stores,	at 7s. . .	4,599	0	0
60	Conductors,	at 2s. .	2,190	0	0
115			£13,996	15	0*

With regard to the employments of Storekeepers, whose duties are of course much the same as the preceding, the Committee on the Army and Ordnance Expenditure recommended, at least partially, a similar change:—

"Your Committee took some evidence regarding the duties of storekeepers, with a view of ascertaining whether retired officers and non-commissioned officers of Artillery might in many cases be competent to undertake the duties of storekeepers and clerks. Your Committee are anxious to call attention to this subject, because they believe that, if retired officers and non-commissioned officers were judiciously selected for these appointments, some saving might be effected for the public, and that the measure might be rendered conducive to the efficiency of the Artillery.

and non-commissioned officers in this service.

"Artillery officers are found to be competent to superintend the manufacturing departments at Woolwich; they are necessarily rendered acquainted with a great variety of stores during

at once great clearness of arrangement and a correct knowledge of the details of ordnance ammunition and stores.—Remarks on the Organization of the Corps of Artillery, &c., p. 126.

* Ibid. p. 159.

their service at home and abroad. Your Committee, therefore, are disposed to think that retired officers of Artillery might occasionally be selected to fill the situations of storekeeper, and it is believed that there are many who would gladly avail themselves of such an offer."*

With regard to the changes proposed in the Artillery service by Mr. Hume, supposing the unit in this case to be a battery as fixed in 1838,† it would only require the addition of twenty-one gunners and two officers to form such a section of artillery under a Lieut.-Colonel, as that which has been proposed to man eight guns. This would consequently be a step in the right direction, and so far an improvement on the existing company system, that there would be a larger portion of the service together, with a fair proportion of company and field officers. But it would, at the same time, have the disadvantage of being deficient in the higher organization of battalions and superior officers, so essential to the well-being of the service. Failing, however, the accomplishment of the latter object, the preceding, or some similar arrangement, could not but be advantageous to the regiment. If carried into operation, it would leave to the Master-General and Board the management of the extensive duties belonging to the barrack de-

Sections of artillery under lieut.-colonels

would be an improvement, though

still wanting in the higher organization.

* Second Report of the Select Committee on Army and Ordnance Expenditure, 1849, p. 38.

† Report of the Commissioners with regard to Naval and Military Promotion, &c., 1840, p. 82.

partment, as well as of the vast material necessary for land and sea service. The military branches meanwhile would become an integral part of the army, and eventually have their strength regulated by, and bear a due proportion to, that of the army at large; whereas, hitherto, the number of these two corps has been affected, if not regulated, by the amount of the annual Ordnance Estimates as a whole. Moreover, it can be shown from unquestionable authority, that even at the height of the war, when England had everything at stake, the Artillery service was sadly deficient. The Duke of Wellington thus expresses himself:—*

Advantages of separating the military branches of the Ordnance expenditure.

The Duke of Wellington complains

"I am concerned to find that I should have made any requisition, a compliance with which is likely to be injurious to the interests of the country, and to interfere materially with the home service. If it is so, the Ordnance establishments of the country, however high, are *too low for the strength of the army;* as the equipment of ordnance stated in my Despatch of the 18th December, 1812, is infinitely lower than that of any army now acting in Europe, of the strength of the British part of the allied army under my command alone, and below the scale which I have ever read of for an army of such numbers."

of the insufficiency of the field artillery.

Other disadvantages of a separate system.

But the continued separation of the services brings many other evils in its train, among which may be mentioned the *delicate* position of the *respective* officers, when (as a matter of duty)

* Letter to Lord Bathurst, dated Frenada, January 27, 1813, in Gurwood's Despatches.

reporting the proceedings of the General commanding, or of the Governor of a colony, to the home authorities, as was the case at Hong Kong in 1846.

Painful as it must be to revert to such circumstances, it becomes necessary to overlook personal considerations, in order to show some of the disadvantages arising to the service from separate and conflicting duties.

The respective officers at Hong Kong.

The writer, who commanded the artillery in China, was, as the senior ordnance officer, President of the Board of Respective Officers, which as usual consisted, in addition to himself, of the commanding royal engineer (Major Aldrich), the ordnance storekeeper (Mr. Pett), and the deputy ordnance storekeeper (Mr. Boate). Acting under direct orders from the Master-General and Board of Ordnance, they had various duties to perform relative to the barracks, the consumption of stores of all kinds, and in fact, to a large portion of the expenditure. The matter that subsequently became so important had reference to the removal of stores, which, contrary to the custom of the service, took place from one part of the town to another by order of Major-General D'Aguilar, the officer commanding in China; and a statement was made to the home authorities accordingly.

Report on the removal of stores.

Major-General D'Aguilar having learnt from one of the respective officers, who had dissented from this proceeding, that such a report had been forwarded to the Board of Ordnance, ordered it to be produced, contrary to what is laid down in the Ordnance regulations, as well in those of Her Majesty;* and having, in the first instance, assembled certain officers to examine into the case, the following was printed and promulgated:—

General order issued by Major-General D'Aguilar,

"GENERAL ORDER by Hon. Major-General D'Aguilar, C.B.,
"Commanding the Troops in China.

"*Head Quarters, Victoria, Hong Kong,*
"*23rd September*, 1846.

"THE systematic opposition so long exercised against the Major-General's authority by Brigadier Chesney and Mr. Ordnance Storekeeper Pett, in their character of respective officers, and under the plea of the Ordnance regulations, has at length manifested itself to a degree that renders it necessary for the Major-General to perform a painful duty to the service as well as to himself.

"Those officers have been repeatedly warned by the Major-General, that he would admit of no infringement of the regulation contained in the Ordnance Circular of the 16th December, 1839, and more fully enforced by another letter from the Board of Ordnance so late as the 5th November, 1845, which requires that all communications from the respective officers, affecting the acts or decisions of the officer commanding at the station, should be made known to him previous to their transmission to England, in order that his opinion may be exhibited in conjunction with their own.

"In the face of this regulation, as well as of the Major-General's repeated commands, Brigadier Chesney and Mr. Ord-

* See the Duke of Wellington's Letter, pp. 250, 251, of the Queen's Regulations; third edition.

nance Storekeeper Pett addressed a letter to the Secretary of the Board of Ordnance on the 22nd June last, the obvious tendency of which was to impress the Home authorities with a belief that the Major-General, by directing the removal of the Ordnance stores from one part of the town to another at that season of the year, had subjected those stores to incalculable damage.

"It might have been expected that a military officer of Brigadier Chesney's standing in the service, and a civil officer of Mr. Pett's presumed experience, would not have drawn up a report, and that too in opposition to the recorded dissents of the Commanding Royal Engineer and Deputy Ordnance Storekeeper, so calculated to alarm the Board of Ordnance (at the Major-General's expense) for the safety of the stores, without having some strong facts upon which to ground their assertion.

on the 23rd September, 1846,

"The subjoined opinion of a Board of Officers, convened by the Major-General to ascertain the real state of the stores, will show how far Brigadier Chesney and Mr. Ordnance Storekeeper Pett were justified in the statement they thought proper to make:—

"'*Opinion.*—As regards the statement put forth by Brigadier Chesney and Mr. Pett in the Report of 22nd June last to the address of the Secretary to the Board of Ordnance, viz.,—"No extensive removal of stores should have taken place at this particular period (the height of the rainy season): the whole of the most valuable stores at the station having been removed in open boats during a ten days' incessant tropical rain, the ultimate deterioration of the stores from this cause can scarcely be estimated,"—the Board are of opinion, that no damage whatever has been occasioned by the removal of the stores in question, nor is any to be apprehended. They are further of opinion, that the boats used were not open, but covered with tilts; and that the rain was not incessant for ten days, but only continuous rain for one working day, 15th June, regarding two of the respective officers.

and no rain whatever on the 8th, 12th, 13th, 18th, 20th, 23rd, and 24th June. Under these circumstances, they regret to be obliged to give it as their opinion, that Brigadier Chesney and Mr. Pett had not any just or reasonable grounds for making the statement they did in their Report of 22nd June last to the address of the Secretary of the Ordnance.

"'The Board have had occasion particularly to remark the efforts used by the Ordnance Storekeeper to construe the rules of his own branch of the service and his duty to the Ordnance Board into an excuse for an act of contumely towards the Honourable the Major-General Commanding, and disregard of the principles of subordination, carrying it even to the preposterous length of maintaining that, while restrained by these rules from taking means to ascertain whether the stores were really damaged or not, the respective officers were justified, and deemed it in every respect a better course at once to accuse the Honourable the Major-General Commanding of having, by his order for the removal of the stores, caused them to be damaged to an extent that can scarcely be estimated, as expressed in their Report, dated 22nd June, 1846.

"'The Board conceive it impossible that either the public interest or the good of the service can have any association with views of public duty so mischievous and unreasonable, or that cordial co-operation in the real objects of the service can be expected while such views are suffered to be entertained.

"'THOMAS SCOTT REIGNOLDS,
Lieut.-Col. 18 R.I. Regiment, *President.*

W. MILLER,
Dep. Commissary-General,
EDW. W. DURNFORD,
Capt. Royal Engineers,
} *Members.*'

"The Major-General entirely concurs in the above opinion, because he never can admit for a moment that the Ordnance regulations are founded upon any principles but those that secure the performance of its departmental duties in accordance with military discipline and the acknowledged rules of propriety; and because he feels that he is evincing his best respect for those regulations by enlisting them on the side of good order and subordination.

"The Major-General is willing to believe that Brigadier Chesney, however mistaken in his views, and culpable in giving effect to them without the Major-General's knowledge, has been led by the representations of the Ordnance Storekeeper to become a participator in a statement now declared to be directly at variance with the facts, and with the real merits of which he was, by his own admission before the Board, totally unacquainted. But, under all the above circumstances, the Major-General has decided upon at once dispensing with the further services in this command of both those officers.

Brigadier Chesney removed from his command,

"Brigadier Chesney will, accordingly, make over from this date the command of the Artillery in China to Lieutenant-Colonel Brereton (who, under ordinary circumstances, would have relieved him next month), and will return to England under instructions that will be communicated to him by the Assistant Adjutant-General. Mr. Ordnance Storekeeper Pett is suspended from office by the Major-General until the pleasure of the Master-General upon his conduct be known.

and Mr. Pett suspended from his duties.

"A remain of the Ordnance stores will be duly taken, and proper measures adopted for their preservation and security.

"By order of the Hon. Major-General Commanding,

"(Signed) J. BRUCE, Captain,
"Assistant Adjutant-General."

In justice to one of the parties, the two words "being absent" should have been added to the "admission of being totally unacquainted," &c., for the Board of Officers were informed, and Ge-

When the stores were removed, Brigadier Chesney was absent.

neral D'Aguilar was perfectly aware, that the writer was in the north of China when the removal of the stores took place. But let us pass from what is in some degree personal, to the more general question.

Appeal made to the home authorities

The officers thus summarily dealt with, lost no time in appealing to higher authority, explaining, in the first place, that having previously objected to the unusual proceeding on the part of the Major-General of ordering the removal of the Ordnance stores, they had, after this had been effected in the rainy season, reported the circumstances to the Board, more particularly as the Storekeeper, not the Major-General, was responsible for their safety. It was also shown from the official register kept at the Government offices in Hong Kong, not only that there was rain on certain days when the stores were being removed, but that during the month of June, 1846, a quantity had fallen nearly equal to the average of the whole year in England. With regard to the nature of the boats, these, according to an official statement from a naval officer, the harbour-master of Victoria, were not covered in the ordinary sense of cargo boats, but merely provided with bamboo matting; which, irrespective of the distance between the boats and the storehouses, was not considered by merchants a sufficient protection for their goods during heavy rain. Besides the questions of the rain, and the nature of the boats, neither of which was examined by the Board of Officers in the presence of the parties, they found that evidence had been taken on other matters during their absence, nor were they even made aware, directly or indirectly, that an opinion was to be given on their conduct; which opinion, one of the Board *being a civilian,* was, it is believed, contrary to the received practice of the service when a military officer is concerned. Added to this, the storemen, who had so much to do in conveying the stores from and to the boats, were not examined on this occasion at all. Moreover, the request of the respective officers to be furnished with a copy of the proceedings was denied, and the Indian pay, &c., of Mr. Pett, and subsequently that of the writer, was stopped by the order of Major-General D'Aguilar.

Under these circumstances both officers not only solicited, but implored the Master-General to obtain a hearing before a competent tribunal, in order that those accusations might be met, which had already been extensively circulated to their disadvantage in the various newspapers of both hemispheres.

for a public hearing of the case.

In conformity to the standing orders of the Master-General, that questions arising abroad should be settled on the spot,* the respective

* Letter of the Dep. Adjutant-General, Royal Artillery, Woolwich, February 27, 1845.

officers remained for six months at Hong Kong, suspended from pay as well as from duty, awaiting a reply from England.

The expedition to Canton caused

The well-known expedition to the Canton river occurred in this interval; and a circumstance took place in connexion with it, which cannot but show how much even service in the field may be affected by the mischief arising from separate departments.

The writer and a party of Englishmen, desirous of seeing a place then but little known, had proceeded some distance up the Canton river to visit a large, and, as it proved, a rich and handsome manufacturing place called Fuchan, or Foot's Town. Here an attempt was made (as Keying expressed it) by "myriads of people" to stone the visitors to death. The author does not attempt to describe the difficulties which the party met with in passing through the town,* nor the still greater dangers which they encountered as their boat passed along the narrow river, exposed to the stones thrown from both sides by the infuriated populace. As such an account would necessarily involve a lengthened narrative, it will be sufficient to say, that although no individual was altogether uninjured, and some were grievously and dangerously wounded, the whole party

by an attack made on the author by the people of Fuchan.

* As this occupied thirty-five minutes, and the streets were closely lined on both sides, the number of people in the town alone probably exceeded 19,000.

escaped the determined purpose of stoning them to death; this fate, after an hour and a half of extraordinary peril, was, under Providence, averted, owing to the noble conduct of a mandarin, and to the Chinese boatmen, who generously exposed their own lives to endeavour to save those of the strangers.

A demand of satisfaction for so atrocious an attack was made, but, as in a previous similar case, evaded by the viceroy Keying. Sir John Davis, as Her Majesty's representative, thought it right to take measures to cause his remonstrance to be respected. He therefore sent for the writer, and having made some inquiry respecting the state of the Bogue forts, decided that an expedition should start towards Canton the same night, and expressed his desire that the writer should accompany it. As there were but two other artillery officers at that time on the spot, the writer thought it his duty to offer his services to Lieutenant-Colonel Brereton (his junior in the corps but senior in the army), which were accepted, and he was with the armament in consequence. But Major-General D'Aguilar coming on board the Honourable Company's steamer "Pluto," desired him to quit the vessel, and the writer found himself in the unenviable position of being turned adrift in a small boat, with one man, in the Canton river, and left to take his chance, when the

The author accompanies the force towards Canton,

as a volunteer, and is sent away by Major-General D'Aguilar.

hostility of the Chinese was naturally excited by that part of the attack which followed the spiking of the guns of the Bogue forts. This took place after he had personally endeavoured to soften the matter by reminding the general that "this was service;" but as the latter persevered notwithstanding, observing that "having reported the writer to the Duke of Wellington, he would not permit him to remain," we must, considering how very much the well-known disposition of this officer was ordinarily opposed to such an extreme step, look for some explanation of the circumstance in his belief that the respective officers at Hong Kong had purposely availed themselves of the Ordnance regulations to give him annoyance, and injure the service at the same time. Whereas nothing could have been farther from their intention.*

Cause of this unusual proceeding.

But to return to the subject itself. A letter reached Hong Kong from the Board of Ordnance about this period, which, referring to a previous occasion, when Major-General D'Aguilar had called for another report made by the respective officers to the home authorities, contained the following:†—

The reports of the respective officers should not

* The ulterior object of sending an expedition up the river towards Fuchan being still unaccomplished, the author continued on the spot agreeably to the intention of Sir John Davis, until the demands made on the people of that town were acceded to, and further proceedings became unnecessary.

† Letter to the Respective Officers, No. 53, par. 9, Office of Ordnance, October 12, 1846, O. 145.

"The Master-General and Board avail themselves of this opportunity of referring you to the circular letter of his Grace the Duke of Wellington, dated 31st December, 1827, addressed to the officers commanding the troops at foreign stations,* on the subject of such officers calling upon Ordnance officers for the production of their reports of proceedings made to the Master-General and Board; and although the question which led to that circular originated in a dispute with a Barrack-Master, yet it very clearly shows his Grace's views of the rules of the service, and the necessity for those reports being sent home direct to the Master-General and Board, as it cannot be expected that these reports will contain the officers' real views of the transaction to which they relate, if they are liable to be called for by any authority whatever."

be called for by the general officers.

Considering the view taken in this passage, and the tenor of the rest of that document, it was rather unexpected to find subsequently† that the Master-General and Board had ordered the writer to return to England, and Mr. Pett to take charge of the duty at Colombo. No reply was, however, given to the appeal which they had made for a court-martial, as the best means of doing justice by ascertaining how far the respective officers had been to blame in this affair.

The author is ordered to England, and the storekeeper to go to Colombo.

Whether any, or what communication may have been received by Major-General D'Aguilar from the Horse Guards in reply to the appeal so frequently and so urgently made for a hearing before a court-martial, is still unknown to the parties concerned; but the writer was thus an-

Uncertain whether there was any reply from the Horse Guards.

* See Queen's Regulations, pp. 250, 251.

† Letter of January 19, 1847, M. 19.

swered by the Master-General of the Ordnance:—

Communication of the Master-General to the author.

"*Deputy Adjutant-General's Office,*
"*Woolwich,* 13*th March,* 1848.

"SIR,

"MAJOR-GENERAL PATERSON having forwarded to me your letter of the 4th instant, together with its enclosures, I submitted them to the Master-General, upon which his Lordship has made the following minute:—

"'As far as I, as Master-General of the Ordnance, am concerned, I consider the case of Lieutenant-Colonel Chesney to be closed.

"'I have shown that no evil impression is made upon my mind respecting Lieutenant-Colonel Chesney, by having placed him in a responsible command soon after his return from Hong-Kong, and I cannot interfere further in the matter in question.'

"I have the honour to be, Sir,
"Your most obedient Servant,
"H. D. ROSS, D. A. G.

"*Lieut.-Colonel Chesney,*
"*Commanding Royal Artillery, Ballincollig.*"

Necessity of some change concerning the respective officers.

Although painful in many ways, it seems but right to show without reserve the serious position in which the respective officers were placed at Hong Kong, in the hope that it may be the means of preventing a similar state of things in future, by putting an end to the objectionable regulation of their being obliged, under certain circumstances, to report against a superior officer, such as the governor or general commanding a colony.

The very unusual order issued by Major-General D'Aguilar was, however, only the result of the last portion which had been added to the previous burthen, or, as he terms it, "systematic opposition."

Although long impressed with the very strongest belief that exclusive duties and separate services were fraught with serious evils, the author felt it to be right to carry out the duties in strict conformity with the regulations of the Board of Ordnance, notwithstanding the difficulties inseparable from an infant colony.

Difficulties attending an infant colony.

Major-General D'Aguilar considered that the health of the troops justified an extensive outlay as "*emergent services*," that is, without the usual sanction; but the authorities in England appeared to consider that a departure from the ordinary course should be confined to the extreme cases of war or rebellion; and in a letter from the Board of Ordnance of the 18th March, 1846, $\frac{11}{88}$, colonial expenditure was strictly forbidden, without previous reference from the respective officers to the authorities in England. In consequence of this communication, it was shown by the respective officers* that the heavy outlay in question had been incurred by the commanding engineer, on the authority of the major-general, without their intervention.

Supposed neglect of duty by the respective officers.

Irrespective of diminishing the outlay by adopting a plainer style of architecture, and by other

* Report to the Board, 27 to 30, August, 1846.

Major-General D'Aguilar believed that he was purposely thwarted;

means, it cannot be doubted that the course pursued by the respective officers on this and several occasions, had the appearance of thwarting those plans which the general thought desirable for the improvement of the colony, and could not therefore be easily borne by a high-spirited commander. Were it, however, yet to be done, it is not improbable that Lieutenant-General D'Aguilar would, on the recurrence of a similar case, recollect that he was pronouncing judgment in his own cause; and as those whom he treated so unceremoniously could neither explain to their friends, nor obtain a hearing for several months, he would not have printed and circulated such an order before the case had been submitted to higher authority, or at least to some competent tribunal.

whilst the respective officers believed they were performing their duty.

On the other hand, the respective officers believed, as they still do, that they simply performed their duty, and that "it cannot be expected that these reports will contain the officers' real views of the transactions to which they relate, if they are liable to be called for by any authority whatever."* But even if the spirit of the latter order had been carried out, first granting an investigation, and then, *if deserved,* giving due support to the respective officers at Hong Kong, the serious objection of junior officers reporting upon their commanders would still remain in its full force,

* Queen's Regulations, p. 251, third edition.

till remedied, as it is hoped it will be eventually, in connexion with other ameliorations in our service.

Evils arising from tardy promotion,

It cannot be denied that the slowness, if not the paralysing decrepitude, of worn-out frames, belong to the superior officers of artillery. Till these evils can be remedied, persons arrived at a certain time of life, as well, indeed, as their successors, must relinquish all thoughts of obtaining that confidence from the general commanding in the field (probably their junior in years) on which the opportune and efficient employment of the arm depends.

to be remedied during peace.

Therefore, while England enjoys the blessings of peace, let her provide a timely remedy, rather than oblige her general commanding, in the hour of emergency, to send home a succession of commanding officers, as was the case in the earlier part of the Peninsular war, until at length the command of the artillery fell to an energetic captain, then the fortieth of his rank.* Let us also hope that the long-standing anomaly of a separate service will cease at the same moment of brightening hope.

Discontinuance of separate services.

Experience has fully proved that in war every operation, whether naval or military, from the smallest to the greatest, ought to be under the absolute direction of one mind, to which every

* Major-General Sir Alexander Dickson's Evidence before the Commissioners for Inquiry into Naval and Military Promotion, &c., 1840, No. 801, p. 48.

inferior officer should be subjected without any reserve on his part, or any possible counteraction by any other department.

Its probable advantages.

With reference to the authority of the Commander-in-Chief, it may be asked whether, if the Ordnance military corps had been equally subject to the Commander-in-Chief, there would at this moment be 30,497 Queen's troops serving in India without any of Her Majesty's artillery at all? Also, whether such a palpable deficiency of artillery would have been for a moment permitted, as that complained of by the Duke of Wellington, in his letter of the 27th February, 1813, from Frenada?* Or that, to which His Grace called the attention of the Government of the country, at a later period, in the following remarkable words:—

"*Bruxelles, April* 15, 1815.

"My Dear Lord,

"I assure your Lordship that the demand which I have made of field artillery is excessively small. The Prussian corps on the Meuse, of 40,000 men, has with it 200 pieces of cannon; and you will see by reference to Prince Hardenberg's return of the Prussian army that they take into the field nearly 80 batteries manned by 10,000 artillerymen. These batteries are of 8 guns each, so that they will have about 600 pieces. They do not take this number for show, or amusement. And, although it is impossible to grant my demand, I hope it will be admitted to be small.

"Yours, &c.,

"*To the Earl Bathurst.*" "Wellington.

---

* See page 216.

Indeed, it may, on the contrary, be confidently affirmed that if the artillery had been equally subject to the Commander-in-Chief, the great pains bestowed on the service by the Duke of York, would have secured for the armies in question, as well as for every other field force, a due proportion of artillery. Nor is it going too far to express a confident belief that whenever the British force shall become essentially one army, it will, whether consisting of 50,000 or 150,000 men, have at all times that proportion of artillery which may be fixed upon as suitable for the wants of the country, abroad as well as at home.

British troops to become one army,

In this case, and as the consequence of incorporating the Ordnance corps with the army, not only will the regiments of the line have the advantage of being exercised with a proportion of artillery, but some of this arm will be permanently attached to the household troops; and, as in other countries, the cavalry and infantry of the Guards will attend the Sovereign accompanied, as the case may be, either by horse or field artillery.

by incorporating the Ordnance corps.

The recent Report from the Select Committee on the Army and Ordnance, of the 21st July, 1851, without speaking very decidedly, and quoting from Earl Grey,—

"Is of opinion that the authority for directing the measures which are required for the effective and economical conduct of military business is divided among too many officers, totally separate and independent of each other, and that this inconvenient division of authority practically prevents arrangements working so well as they ought.

Consolidation of departments

"That it is impossible for the Colonial Office to exercise a constant superintendence over all the arrangements with regard to the military force of the country, in consequence of the greatly-increased duties of the Secretary of State for the Colonies since the last war.

"He is of opinion, that if the recommendations contained in the Report of 1837 were carried out, it would lead both to greater efficiency and more economy in the administration of the army; and these suggestions might be carried into effect without interfering with the military command of the officer at the head of the army."

under a War Minister.

These passages at least imply, that in a widely-extended empire like that of Great Britain, with an army much divided, and, if India be included, composed of various races, the special and immediate control of a War Minister is most desirable, in order to remedy the *delays* and disadvantages of separate departments, which would thus be replaced by the speed, efficiency, and economy of the general centralization of the various departments and services.

CHAPTER VIII.

WANT OF VITALITY IN THE ARTILLERY SERVICE.

Disadvantages of Promotion by Seniority.—Occasional Promotion for Merit desirable.—Difficulty of Deciding upon Merit.—Conflicting Opinions regarding Promotion out of turn.—Necessity of this change illustrated by the Peninsular War.—Promotions out of turn by Lord Loudon, Sir Jeffrey Amherst, and the Earl of Albermarle.—General Officers to be selected for employment.—Opinion of Earl Grey.—Talent and exertion gain distinction at the Academy, but receive no stimulus in the Regiment.—Instruction of the Second Lieutenants superintended by a Captain.—Cadets in the Arsenal to have the rank of Second Lieutenants.—The British Artillery does not serve in India.—Disadvantages of this exclusion.—British Artillery in India would cause a spirit of emulation.—The East India Company's Artillery varies at the different Presidencies.—Captain Oakes' proposition for an uniform organization of the Indian Artillery.—Uniformity urged by the Court of Directors.—Employment of British Artillery in India would effect this object.—Infantry Drilled at the Guns instead of increasing the Artillery.—Disadvantages of this system.—Present strength of the Artillery.—Proposed strength of the Regiment.—Number and strength of the Companies.—Proportion of Officers.—Present and proposed Numbers of Officers and Men.—Horse Brigade.—Its present and proposed strength.—Effects of the proposed change.—Colonels and Lieut.-Colonels to become one rank in the Line and Artillery.—Advantages of this change to the Service.—The present project gives more Officers in proportion to the Men than has hitherto been the case.—Separation of the Arm into Field and Garrison Services.—Doing Duty by Battalions.

Disadvantages of promotion by seniority.

RETURNING from this digression to another question, advancement by seniority—the certainty of promotion in turn, whether the individual be well or ill qualified, has the great disadvantage of taking away the strongest incentive to exertion, namely, the prospect of acquiring distinction. For this reason exceptions should undoubtedly be made in favour of the somewhat rare instances of brilliant and distinguished service; and also of professional qualifications and attainments of the very highest order. The bare possibility of promotion in this way, were it only bestowed on one out of each successive ten at the top of the list, must speedily change a state of hopeless apathy for one of continued emulation, not only in the pursuit of theoretical knowledge, but also in following this up by animated exertions to obtain excellence in the ordinary practical duties of an officer.

One step in ten proposed to be given for merit.

It has with justice been observed, in reply to what was shortly stated in the former publication, that—

Difficulty in deciding on relative merit.

"It would be difficult to measure and contrast with any degree of accuracy the relative merits of individuals scattered over the whole world, and who, though unknown, because at a distance, may be superior to those who are to decide upon their relative qualifications. Such a change as promotion out of turn would, it is alleged, open a wide door to favouritism and abuse the most cruel, since an officer without spot or blemish may be wounded and mortified whilst doing his best for the welfare of the service according to the humbler gifts bestowed on him by nature."

Objections such as these were not by any means overlooked, and the writer freely admits their force and justice; but there are other, and perhaps weightier, considerations to be taken into account. It will be seen that the officers who were examined on this question* differed in opinion about the advisability of promotion out of turn. Colonel Cator, of the Royal Artillery, gave it as his opinion that promotion by seniority, as far as regards the efficiency of the service, is a very bad system;† and the late Lieutenant-Colonel E. Michell thought that some change in this respect was necessary.‡ Lieutenant-General Thackeray, R.E., with reference to its disadvantages, recommended exchanges of engineer officers to and from the line.§ Major-General Sir Charles Pasley considered that an opening for merit, if it could be judiciously done, would create a great stimulus in the engineer corps; || and he also stated that it had been proposed to promote one out of six of the subaltern officers of the corps out of the routine of seniority.¶ Lieutenant-Colonel Reid, R.E., thought that brevet or regimental promotion for distinguished services in the field, would cause very great satisfaction in the corps; that no dissatisfaction or jealousy could arise if it were the rule, and that it would be beneficial to the service.** Everything, he also remarked, that creates emulation is desirable; and that such promotions should not be confined to the field.†† Lieut.-

Opinions in favour of promotion out of turn,

and probable benefit to the service.

* See Report of the Commissioners for Inquiry into Naval and Military Promotion, &c., 1840.

† Ibid., No. 684.

‡ Ibid., Nos. 757, 763.

§ Ibid., Nos. 946, 947.

|| Ibid., No. 1107.

¶ Ibid., Nos. 1118, 1136.

** Ibid., Nos. 1270-1273.

†† Ibid., Nos. 275, 1763.

Colonel Matson suggested, amongst other ameliorations, that one-tenth of the promotions from subalterns should be by selection for merit, to be decided upon by a competent Board;* which, the merits of officers being known, would not be difficult; and thus a stimulus being excited, it would be a benefit to the corps.†

Sir Alexander Dickson's objections to such change

But on the other side, General Fanshawe considered that there would be great difficulty in working out an alteration in the seniority system without injustice;‡ and a still higher authority took the same view. The late Major-General Sir Alexander Dickson was very adverse to impugning the principle of seniority, in the belief that the change would be neither good for the public service nor for private feeling.§ Elsewhere, alluding to regimental rank out of turn, he says—"I see great difficulty in it; I think it would lead to an interest that might be exercised where officers might be brought forward without its being quite by their own exertions."‖ But to the same question put elsewhere he replied, "I do not object; but the cases should be of very peculiar distinction."¶

Doubtless influence and favouritism will operate to the end of time, to the extent at least of giving,

* Report of the Commissioners for Inquiry into Naval and Military Promotion, &c., No. 1301.

† Ibid., Nos. 1306, 1318, 1319, 1320, 1321, 1322.

‡ Ibid., No. 1221.

§ Ibid., Nos. 877, 878.

‖ Ibid., No. 842.

¶ Ibid., No. 840.

in some instances, a preference to those who are barely equal, or it may be, even inferior in point of claims to other candidates. Still it is unlikely that a departure from strict justice will be carried farther; and it is to be expected that departure from promotion by seniority will be based either upon professional attainments of the very highest order, such as those of the late General Shrapnell, or, to use the words of the Commissioners of 1832, upon "such pre-eminently distinguished services as should place the individual above all rules;" and thus, as it may be added, carry, as it would have done in the instances of Sir Augustus Fraser and Sir Alexander Dickson, the support of the whole corps.

would be counter-balanced by many advantages, and

The modesty which ever accompanies real merit, and his affection for the artillery service, caused the latter officer to forget that his own case shows, beyond all question, the necessity of being able to depart occasionally from the strict rule of seniority. His rank in the Portuguese service enabled him to be placed in command of the allied artillery, although at that time only fortieth in the list of captains in our service; and had it not been for the former accidental circumstance, he must have been advanced out of turn, or the Duke of Wellington would have been deprived of one of his ablest supporters throughout the arduous campaigns in the Peninsula.

his former position shows the necessity of this change.

Necessity for some stimulus in the service.

But it was the necessity for some kind of stimulus which might infuse life into a service, in this respect to be considered as dead, that caused the writer to suggest the possibility of promotion out of turn, to the limited extent, however, of one out of ten at the top of each grade. Nor is advancement out of turn on special occasions unknown in our service. Certain officers of artillery were promoted in this way by the Earl of Loudon in Portugal, by Sir Jeffrey Amherst in North America; and this was also the case with Majors Hislop and Maitland in the East Indies.* The same thing was done more extensively by Lord Albemarle as a reward for the continued and trying service in taking the Havannah, 14th August, 1762, when the following promotions were made by his Lordship:—

Lord Albemarle's promotions.

"Capt. Lieut. G. Lewis to be Capt., *vice* Stanley, killed.
1st Lieut. Wm. Lee to be Capt., *vice* Strachy, killed.
2nd Lieut. Pattison to be 1st Lieut., *vice* Lee.
Lieut. Fire-worker J. Brietche to be 2nd Lieut., *vice* Pattison.
Cadet J. Bossom to be Lieut. Fire-worker, *vice* Brietche.
Lieut.-Col. S. Cleaveland to be Col., *vice* Leath, dead.
2nd Lieut. J. Lemoine to be 1st Lieut., *vice* Benjamin, dead.
Lieut. Fire-worker J. Bloomfield to be 2nd Lieut., *vice* Lemoine.
Sergt.-Major Wm. Grant to be Fire-worker, *vice* Bloomfield.

---

* Kane's List of the Officers of the Royal Regiment of Artillery Advertisement.

Mr. J. Reeves to be Lieut. Fire-worker, *vice* Bossom, resigned.

Lieut. Fire-worker Thos. Brady to be 2nd Lieut., *vice* Brietche, resigned.

Sergt. J. Hooke to be Lieut. Fire-worker, *vice* Brady."*

These Commissions were confirmed by the King; but what is now proposed would be much more limited in its operation, since only nine places could be obtained; namely, when an individual who has already merited distinction arrives within ten places of the top of his present rank, and then passes over nine of his seniors.

Its probable effect on artillery officers.

But perhaps the question may be best viewed as a choice between the evils of inertness on the one hand, and the incentive of bright and animated hopes on the other. The latter, it is true, would be attended with serious evils; but these, it is to be hoped, would be more than counterbalanced by considering the interests of the nation rather than those of individuals, a distinction which has been so happily expressed elsewhere.

According to Earl Grey, officers should be selected by merit.

Alluding to the advisability, if not the absolute necessity, of selecting general officers for promotion with a view to employment whilst in the prime of life, Earl Grey, on being asked by the Committee of the House of Commons† whether

* From the Artillery Records, collected by the late Colonel Cleaveland, R.H.A.

† Report on the Army and Ordnance Expenditure, August 1, 1850, p. 736, No. 9392.

any system of selection would not be invidious, and open to the charge of favouritism, replied, "I think it would; but I think that though this is an evil, it is a much lesser evil than what now exists."

An impetus is given in the earlier part of the career of an artillery officer which unfortunately is not subsequently maintained. Sir Augustus Fraser observes:—

Talent and exertion give advancement at the Academy.

"Whilst at the Academy, advancement depends upon exertion in the several branches of study; and, hence, an emulation is excited most favourable to the development of talent and ability.

"On joining the corps as officers, the young men are required to remain for some time, generally for a year or more, at the head-quarters of the regiment at Woolwich, for the purpose, as it is said, of acquiring some knowledge of their profession before they join the companies to which they are posted. But as there is no officer whose duty it is to instruct these young men, or to be in any way responsible for their improvement, and as there are generally at Woolwich a great many of these young officers, between whom and the senior officers of the regiment much intercourse, from disparity of age, is not likely to exist, they find themselves obliged to associate almost exclusively with each other; in many cases gradually lose much of their earnest desire to obtain professional knowledge; and at last join their companies, having already forgotten part of what they may have learnt at the Academy, without having acquired anything in return but the bare habit of mounting guard and hearing the rolls called in the barracks.

The desire for professional knowledge ceases in the regiment from want of stimulus.

"Here, then, is the first error in the employment of officers of artillery, when so much of the success of life depends on the good employment of the first part of it." *

A captain now superintends the studies of junior officers.

This subject has at length attracted attention, and thirty-two years after the appearance of the preceding remarks in print, an officer, whose duties are similar to those of the capitaine-instructeur of the French artillery, has recently been appointed to superintend the studies of the second lieutenants. At some future day the corps will have reason to be grateful to the Master-General and the other authorities for this enlightened step in the right direction, more especially if the grade of second lieutenant should not be considered fixed as to succession, but (as has been the case of late in the Arsenal) should cause those who are most advanced in their studies to be first promoted to companies as lieutenants; these appointments in the mean time being exclusively for instruction, and with a view to competition. Were it established that second lieutenants are not to take permanent positions in the regiment till they leave the captain of instruction, every benefit that can be obtained by talent, assiduity, and competition, would be the result, while a love for study and the habit of application must be the consequences.

The cadets in the Arsenal to have the rank of 2nd lieutenants.

One cause of the backward state of the British

* Remarks on the Organization of the Corps of Artillery in the British Service, p. 8.

artillery has been incidentally noticed already, as having its origin in connexion with our most important colonial possession.

The Queen's artillery, &c. are not employed in India.

Independently of the limited proportion of both services which has hitherto prevailed in the British empire, no artillery or engineers whatever are maintained for the Queen's troops serving in India: for, either owing to the exclusive nature of the Ordnance service, or from some other cause, the British army has been employed in that part of the world without any portion of the Queen's artillery. Had the latter been an integral part of the British army, it may fairly be presumed that the cavalry and infantry would not have been sent to serve in the East, or indeed anywhere else, without a due proportion of the artillery; in which case the East India Company would only have been obliged to raise and maintain a due proportion of this arm for the native service, since the Queen's troops would have had their own artillery.

Diminution of the Royal Artillery in consequence of this exercise of the Directors' patronage.

The additional patronage arising from the present system offers a serious bar to obtaining this boon for the corps, which suffers in consequence; and as long as it exists, the East India Directors will naturally continue to provide the proportion of artillery required for the Queen's troops serving in India, although it is attended with some disadvantages to the empire which they govern so judiciously.

It is well known that the artillery of Europe gains much by the constant attention to the progressive improvement of this branch of military service in the different continental armies; and of such advantage the East India artillery is in a great measure deprived. It cannot, therefore, be any disparagement to those who have performed their duties in the field so admirably, to express the belief that the Ordnance corps in the East would gain as much by the emulation which would be the consequence of the presence of a proportion of European artillery and engineers, as the cavalry and infantry of India undoubtedly do, from having among them a portion of these arms belonging to the Queen's service.

The presence of the Royal Artillery would cause emulation, as in the case of the cavalry and infantry.

The East India Company's artillery has not escaped those differences as to details at the three Presidencies, which, as regards the other two arms, have gradually been giving way to a general system applicable to the whole force. Even the strength of the troops and companies varies in some degree. A golundauz, or bullock battery, for instance, has 88 privates in Bengal, 92 in Madras, and 70 in Bombay. There is also some difference in the numbers of Europeans in a troop of horse artillery, which, including artificers, &c., are 130 in Bengal, 136 in Madras, and 132 in Bombay: an European company has 98, 99, and 123 respectively at these Presidencies. In Bengal

The artillery of the East India Company differs at each of the three Presidencies.

gun lascars are attached to the horse artillery, and also to the European foot artillery, but not to the golundauz; whereas in Madras, the horse artillery is without gun lascars, who are, however, attached both to the foot artillery and golundauz of this Presidency; whilst in that of Bombay they are only attached to the horse artillery and native foot artillery.

The preceding details are taken from a memorandum submitted to the late Commander-in-Chief in India, General Sir Charles Napier, by Captain A. F. Oakes, of the Madras Horse Artillery, who shows that essential differences still exist, not only as to the extent, but also as to the nature of the equipments. For instance, poles, instead of shafts, are used with the gun and limber carriages in Bengal, in which Presidency, instead of mounted detachments, the gunners of the horse artillery are, from motives of economy chiefly, carried on the off-horses of the guns and waggons; thus placing a troop of horse artillery nearly on the footing of an ordinary field battery. It is besides, the practice of the artillery belonging to this Presidency to come into action with the limbers facing the rear, instead of the more rapid system of unlimbering the guns as they advance; which, be it observed, has the advantage of keeping the men nearer to the gun at this particular moment, so that the latter is ready to open its fire by the

The pole used instead of shafts in Bengal,

and gunners mounted on the off horses.

time the horses have wheeled round. Thus it would appear that the artillery at the seat of government is in a more backward state than that of the other two Presidencies; moreover, as stated by Captain Oakes, each government adopts a separate system of drill, differing, too, so widely, that an officer of one establishment would scarcely be able to command and manœuvre a battery according to the practice of either of the other Presidencies, nor would he be able to act with them.

The artillery of Bengal more backward than that of the other Presidencies.

The proposal of Captain Oakes, if adopted, will bring about one general system of drills, manœuvres, and organization, for the artillery of India; so that the 450 pieces of ordnance (138 being horse artillery), fully equipped, as well as the 300 pieces in reserve, and the 15,719 Europeans and gun lascars to man them, may be alike at the three Presidencies in every respect.

Captain Oakes proposes to make the services uniform.

Possibly by this time the plan in question may have been carried out, since as far back as 1838, a committee of experienced artillery officers, assembled from the three Presidencies, reported that a modification of the system of the Royal Artillery ought to be adopted for the whole of the artillery in India.

The disadvantages of want of uniformity appear to have been strongly felt in a higher quarter long before, for the Court of Directors (as quoted in the memorandum of Captain Oakes) thus strongly and pointedly noticed this subject in Par. 23 of their Despatch to the Governor of Bengal, No. 18, July 11, 1834:—

The Court of Directors requires similarity of drills, &c.

"Each Government recommends its own fancied superiority in ordnance improvements, and advocates the suggestions of its artillery officers, little or no attention being paid to what may be already in useful operation at the sister Presidencies, and comparisons which might be beneficial and lead to uniformity are seldom made."

The employment of some British artillery in India would have produced general uniformity.

If the cavalry and infantry serving in India had been accompanied by a due proportion of European artillery, instead of allowing the East India Company to provide this arm entirely, to the manifest disadvantage of the Royal Artillery, the desideratum of uniformity must have been the consequence of the latter going out from time to time to impart to the three Presidencies the latest state of advancement known in Europe.

Infantry drilled at the great guns.

As a temporary means of supplying the well-known deficiency of artillerymen, the practice has for some time prevailed, of instructing a portion of each infantry regiment in the exercise of the great guns. This was particularly the case at Hong Kong, where the effects of the climate had reduced the company of artillery to a few men; and during the short service against Canton in 1847, the writer had the opportunity of seeing those of the 18th regiment, who had been so instructed, performing the duty of artillerymen with real success.

Disadvantages of trusting to this resource,

But admitting that there might be time, as in this instance, to give sufficient instruction to a part of the infantry, the opportunity of obtaining the services of the latter, when most required, is very doubtful; since the officer commanding a regiment, or a ship, could scarcely be expected to compromise the efficiency of either, by sparing his men at a critical moment; and even if he should spare them, the officers to exercise the requisite superintendence would still be wanting. Instead, therefore, of attempting to reinforce the artillery from the line, would it not be preferable, and, as regards efficiency, infinitely better to increase this arm sufficiently to be able to man a portion of the guns in each fortress? and as there would in this case be some men to spare from the artillery duties, they could, especially in time of peace, take a proportion of garrison guards, &c. But as the gunner is more highly paid than the infantry soldier, the increase of this arm in the fortress should be limited to a bare sufficiency to work those guns which may be brought to play upon an enemy attacking it by land or sea; for instance, two fronts in the former case, and the guns that could bear upon a fleet at the same time in the latter.

rather than increasing the artillery.

According to the organization proposed in this

work, the present strength of the regiment, about 10,416 non-commissioned officers and men, would, instead of 96, be divided into 144 smaller companies; each having 1 captain, 2 lieutenants, and 72 artificers, non-commissioned officers, gunners, and drivers. Allotting 6 of these companies, as proposed, to each, there will be 24 battalions or brigades of artillery, with 24 field and company officers; which would cause an annual increase of 859*l.* 13*s.* 3*d.* for the alteration in the grades of the officers, as shown in detail in Schedules 1 and 2.

In the project submitted to the Committee of the House of Commons, a small saving attended the reconstruction of the regiment. The present change has arisen chiefly in consequence of having the addition of 24 quartermasters, principally as an encouragement to the deserving non-commissioned officers, instead of allotting, as before, one officer for the duties of adjutant and quartermaster.

Result of the proposed reorganization.

But as 24 officers would be sufficient for a battalion of 600 men, the regiment may be increased (including the horse artillery) to 15,044 non-commissioned officers and gunners, which is the smallest force that could be considered adequate. This, on the existing system, would have required an augmentation of three battalions, which, as compared to the proposed organization, would be a saving of 29,250*l.* 14*s.* for the wants of the empire abroad and at home, chiefly for the 49 officers whose services would thus be dispensed with.*

The following comparison will show the alterations which would be the consequence of changing the existing 96 into 144 smaller companies of artillery, with three instead of five officers to each, maintaining, however, as nearly as possible, the same relative proportion as heretofore.

According to the Ordnance Estimates,† the Royal Regiment of Artillery has 600 officers‡ and 10,416 non-commissioned officers and men; whose details would be altered from—

At present.	Proposed.
12 Colonels-Commandant	12
12 Major-Generals	12
24 Colonels-en-Second	24
36 Lieut.-Colonels, 1st Class	48
12 Ditto, 2nd Class, on Major's pay	24
96 1st Captains 96 2nd Captains	144
12 Adjutants	24
12 Quarter-Masters	24
192 1st Lieutenants 96 2nd Lieutenants	144 1st Class. 144 2nd Class.
12 Sergeant-Majors	24
12 Quartermaster-Sergeants	24
12 Farriers	12

* See Schedules Nos. 2, 5, and 7.

† For 1850, 1851: February 18, 1850, p. 10.

‡ Exclusive of the chaplains and medical and veterinary officers.

At present.		Proposed.
12	Shoeing-smiths	12
12	Collar-makers	12
12	Wheelers	12
96	Colour-Sergeants	144
288	Sergeants	288
384	Corporals	432
384	Bombadiers	432
192	Trumpeters and Drummers	288
9,000	Gunners	8,736

11,016, or 96 companies. 11,016 to become 144 companies, or 24 battalions of artillery, with 24 officers to each.

The brigade of horse artillery, according to the same estimate, consists of 45 officers and 612 non-commissioned officers, artificers, &c. :—

HORSE BRIGADE.

At present.		To become.
1	Colonel-Commandant	1
1	Major-General	1
2	Colonels-en-Second	1
3	Lieut.-Colonels	2
7	Captains	7
7	2nd ditto	0
1	Adjutant	1
1	Quartermaster	1
21	Lieutenants	14
1	Sergeant-Major	1
1	Quartermaster-Sergeant	1
1	Staff-Sergeant	1
1	Trumpet Major	1
1	Farrier	1
1	Carriage-smith	1
3	Carriage-smiths	3

At present.		To become.
1	Collar-maker	1
14	Troop Staff-Sergeants	7
21	Sergeants	14
21	Corporals	21
14	Bombadiers	14
7	Trumpeters	7
7	Ditto	7
7	Farriers	7
7	Shoeing-smiths	7
7	Collar-makers	7
7	Wheelers	7
354	Gunners	354
136	Drivers	136
612		598

In assimilating this arm to the companies and field batteries, there will be an annual saving of 6,372*l.* 14*s.* 4½*d.*, chiefly for the pay of one colonel, one lieutenant-colonel, seven second captains, and seven lieutenants. The aggregate expense of the artillery service at large would, however, instead of 426,007*l.* 18*s.* 3¾*d.*, as it is at present, be 426,867*l.* 11*s.* 6¾*d.*, in consequence of the other changes,* but giving in return the following promotions:—

8 Lieut.-Colonels to be Colonels en Second.

31 First Captains to be Lieut.-Colonels, 13 of the number on Majors' pay.

103 Second Captains to be First Captains, including the Adjutants, leaving only one Captain to be absorbed.

96 Vacancies for 96 Cadets, to be First Lieutenants on reduced pay.

* See Schedules Nos. 1 and 2.

Twelve additional colonels-en-second, on lieutenant-colonels' pay, were proposed in the former publication, for the double object of making the battalions more efficient at foreign stations, and at the same time afford that outlet for promotion, on which this so largely depends. The latter object might, however, be accomplished, though not so judiciously, by bestowing the rank of full colonel on twelve senior lieutenant-colonels of the regiment, in order sooner to make way for their juniors by their becoming major-generals.

The rank of colonel should be maintained in the artillery

So long as the grade of colonel exists in the army, it is but just that it should be maintained in the Ordnance corps, not only as the means of restoring a part of the rank previously lost by the extreme slowness with which the rank of lieutenant-colonel had been obtained, but also for other reasons, which in a combined service should have great weight. Among these may be mentioned the prominent position of the charge of the artillery service in extensive fortresses abroad,* and the command of battalions of artillery. This ought equally to hold good, whether the battalions shall continue as at present, or on the smaller scale which the author has ventured to suggest as more suitable to the wants of the service.

as long as it is continued in the line.

As it is understood that the necessity has been

* Malta has 486, Greece 351, Canada 309, the West Indies 484, and Ceylon 281 guns. The service is proportionably extensive elsewhere.

both felt and expressed by the highest military authorities in the country that the grade of major-general should be obtained in the army at an earlier age than at present, some means will ere long be found to accomplish this object; and perhaps doing away with either the rank of colonel or lieutenant-colonel in the army might be one of the easiest means of accomplishing so desirable an end.

Admitted necessity of younger major-generals.

The main point, the efficiency of a regiment, does not seem to be changed or affected whether the commanding officer be styled colonel or lieutenant-colonel. In this respect, therefore, one of these ranks might safely be dispensed with, or, which would be the same thing, the two ranks might be united. Were this to be the case, and one list formed of both ranks, the lieutenant-colonels, as is the case with captains of the navy, might arrive at the rank of colonel after three years' service, and they would probably receive the step of major-general in seven to ten years more; that is, from ten to thirteen years from major to major-general, instead of from twenty to twenty-five as at present.*

The colonels and lieutenant-colonels to be united.

Advantages of the change,

A similar change would have equal, if not greater advantages, for the artillery and engineers.

and its benefits to the Ordnance corps.

* The senior colonel of the effective list of the army has now, December, 1851, been ten years in his present rank, and the senior lieut.-colonel ten years in his; and in all probability five years will be added to each before the next general promotion.

Proportion of officers.

The senior officers, whether styled colonels or lieutenant-colonels, would have charge of the battalions and principal commands till the rank of major-general removes them, but of course much sooner than at present, from the effective list.

With respect to the working officers of the Royal service, it has been seen, pp. 90, 91, that the number given in the proposed reorganization is greater in proportion to the men than those who served with a troop or company during the Peninsular war, or those recently serving with this arm in India, and is much greater than that of the artillery service of any continental force, Russia excepted. Moreover, the proposed change to smaller battalions would give (even if financial considerations should prevent any increase of the officers, men, and horses) considerable relief to the corps, and still greater efficiency as the senior officers drop off.

A battalion system may be maintained notwithstanding detached services.

Although considerations of the expense to be caused by such a change cannot be overlooked, a still more important object might be gained by going a little farther, and breaking the regiment into specific branches for field and garrison service, doing duty at the same time, in every instance, by battalions, whether at home or abroad. It is true that some difficulty exists respecting the stations eastward of the Cape; but although they would be more detached and separated in these cases than is desirable, there would be the advantage of concentration when the period of relief arrives. But with regard to nearly the whole of the other stations, both at home and abroad, the unity of a battalion may be as easily preserved as that of a regiment of the line when subject to detachment service, as in Ireland, the West Indies, or other parts of the world.

CHAPTER IX.

ON THE NEW MUSKET.

Importance of the new Musket.—Inventions of the present day.—The new elongated Musket-ball.—Various kinds of Fire-arms.—Breech-loading Muskets and Small Arms.—Experiments with the Norwegian Musket, as given by Colonel Portlock.—First Balls of Delvigne and Lovell.—Tables of the Norwegian Experiments and Results.—Objection to and advantages of the new Musket.—Changes in the form of the new Ball.—Description of the new Prussian Musket.—Its advantages, objections, and remedies proposed.—Sears's Patent Needle Gun, and Lancaster's Rifle.—Consumption of Musket Ammunition in Great Battles, particularly at Waterloo.—The supposed effects of the new Musket in War.—Advantages to be derived from this weapon in Defending Strong Positions.—Colonel Portlock's Observations on the Changes in Fortifications consequent on the new Musket.

THE power of musketry having been much discussed, and indeed greatly increased, particularly by continental nations, since the author's pamphlet appeared in 1849, it seems but right to notice this important question, as one on which the well-being of the arm of artillery, and even the fate of nations, so largely depends.

Inventions of the present day.

This proposed short notice will show that the march of practical science of late years has not been confined to the stupendous structures of tubular bridges, and the power of steam, nor even to making the lightning flights of electricity useful to mankind, but that the laws which regulate projectiles have not only claimed, but obtained, a share of that wonderful progress which distinguishes the present so far beyond every previous period of the world.

An elongated projectile is one of the happier efforts of skill and genius, the application of which to the musket has been accompanied by such improvements in the arm itself, as will, according to some, supersede the use of light artillery altogether; and, under modified circumstances, must produce considerable changes in the formation as well as the tactics of modern armies.

Elongated musket balls, and their probable effects on artillery.

The various kinds of fire-arms now competing for the palm of excellence may be classed under two heads, viz., the breech-loading musket, and another description of weapon receiving the ball at the muzzle. The latter, as far as it has been brought into use in France, is practically shown to be greatly superior to the old musket of that nation, and the former is expected to have still greater advantages; but whichever principle may ultimately receive the preference, it is certain that the new instruments, in either case, will have greater range and far more accuracy than has been hitherto obtained from the best rifles in the hands of the most experienced marksmen.

Various kinds of fire-arms proposed.

Loading at the breech was practised in the reign of Henry VIII., and is still partially in use, both in China and other parts of the East, for

Breech-loading pieces very ancient.

great guns as well as for smaller pieces, particularly jinjals.

Wall pieces and muskets of this kind recently adopted.

Rapidity in firing, and other advantages, has caused this manner of loading to be a desideratum since the time of Marshal Saxe; and the principle was introduced some years ago by the French for wall-pieces, the great length of which, and their importance as defensive weapons in sieges, made such a change very desirable.

These, however, were but partially successful, and it was only recently that the objections raised in 1831 to the fusil de rampart seems to have been overcome by another invention, which promises to obviate the disadvantages connected with the escape of gas through the apertures caused by a moveable breech.

A breech-loading musket used in Norway,

An infantry rifle adapted for loading at the breech was invented in Norway, which was supposed to combine the accuracy of the best rifles with the expeditious loading of the ordinary musket.

The records given by Scharnhorst in his treatise on Danish artillery practice, and various references in the Militär-Zeitung to the proceedings of military commissions in Sweden and Norway, show that the armies of the north are not behind those of more southern nations in attention to the details of military science. Amongst other objects of inquiry, a Commission of officers was appointed in Norway to report upon an infantry rifle loaded at the breech, and extensive experiments were carried on between 1839 and 1845 to test the relative advantages of this weapon compared with the common smooth-barrelled musket, and also with the Jäger rifle, at different distances and in various ways. We are indebted for the results of these experiments to the consideration given by Lieutenant-Colonel Portlock, R.E., F.R.S., to the important question of how far such improved weapons may be made to take a more effectual part in the defence of fortified places and positions; which results he has given in a tabular form in the papers of his corps, 1849–1850.

and compared with the other pieces.

A new musket was constructed under the direction of the Swedish Commissioners, which it was hoped would combine the best qualities of the weapons hitherto in use, and would also be an improvement on the flattened ball invented by Delvigne for his musket, as well as the belted ball introduced in England by Mr. Lovell. The first comparison was made with a rifle brought from Berlin, and considered in point of construction to be the best used in the Prussian army up to that time: both were screwed into a machine, and the results are stated in the following tables:—

First balls of Delvigne and Lovell.

Result of the first experiment.

TABLE I.

Description of Arms.	At 137 Yards.			At 205 Yards.			At 274 Yards.		
	Per centage of Shots striking a Target 6 feet high.								
	Breadth of the Target.								
*	2′	6′	18′	2′	6′	18′	2	6′	18′
Prussian 8-grooved rifle, with $\frac{1}{20}$ ounce charge, the barrel cleaned after 5 shots	88	100	100	68	100	100	36	58	72
Norwegian 6-grooved musket loaded at the chamber, charge $\frac{1}{11}$ ounce, ball 32 to the lb. dipped in tallow, barrel cleaned after 25 shots . .	94	100	100	84	100	100	32	72	78

Two thousand shots were fired from two of these rifles, and as they were found unimpaired, and the wear quite insignificant, the Committee, after several trials, recommended that 400 rifles and 50 pistols should be made, and put into the hands of the soldiers for further trial.†

The King raised the number to 500, and his Majesty attended the experiments made at Aggerhaus on the 8th April, 1845, and a third piece was also used on this occasion, the smooth-barrel musket.

* The breadths of target given here, namely, 2 feet, 6 feet, and 18 feet, made respectively the space occupied by one man, by three men, and by nine men.

† Corps Papers, &c., compiled from the contributions of the Officers of the Royal Engineers, &c., pp. 39, 362–380. John Weale, London, 1849–50.

TABLE II.

TARGET PRACTICE on the ICE at the Fortress of AGGERHAUS, on the 26th March and 8th April, 1845, in presence of the King, with Nine Jäger Rifles and Nine Muskets, both with Flint Locks, and Nine Chamber-loaded Rifles, with Spherical Balls and Percussion Locks. The firing was independent, and continued for ten minutes at each of the established distances.

Subsequent trial in presence of the king of Norway.

Description of Arms.	No. of Shots fired.	No. of Shots striking a Target 6 feet high. At 137 Yards. Breadth of the Target. 2′	6′	18′	At 205 Yards. 2′	6′	18′	No. of Shots fired.
9 Chamber-loaded rifles .	276	93	158	200	27	60	104	231
9 Smooth-barrel muskets .	134	13	33	59	6	10	28	117
9 Jager rifles	80	22	39	50	5	8	12	57

Nine fire-arms of each kind used.

From this experiment it may be assumed that the practice of the chamber-loaded rifle was in ten minutes more effective, by striking the object, than that of either of the other arms, in the following proportion :—

TABLE III.

	In breadth equivalent to about 1 Man.	3 Men.	9 Men.
Than that of the rifle at 137 yards	$4\frac{1}{2}$ times	4 times	4 times
Than that of the rifle at 205 yards	5 ,,	$7\frac{1}{2}$,,	$8\frac{2}{3}$,,
Than that of the smooth-barrel musket at 137 yards . . .	7 ,,	$4\frac{3}{4}$,,	$3\frac{3}{8}$,,
Than that of the smooth-barrel musket at 205 yards . .	$4\frac{1}{2}$,,	6 ,,	$3\frac{5}{7}$,,

TABLE IV.

Description of Arms.	Fired.	Per Centage. Target 6′ high. At 205 Yards. Breadth of the Target. 2′	8′	24′	At 274 Yards. 2′	8′	24′	Fired.
26th March.	Volleys.							Volleys.
10 chamber-loaded muskets fired in line, at command .	10	32	85	90	15	51	69	10
10 smooth-barrelled muskets fired in line, at command .	10	25	56	72	3	18	40	10
10 Jäger rifles, firing independently . No. of hits	Shots.	29	79	84	4	20	31	Shots.
Per centage	100	29	79	84	9	44	69	45
8th April.	Volleys.							Volleys.
10 chamber-loaded muskets firing in line, at command .	10	29	60	75	17	42	71	10
10 smooth-barrelled muskets firing in line, at command .	10	4	25	52	..	..	..	..
10 Jäger rifles, firing independently . No. of hits	Shots.	10	25	40	8	26	37	Shots.
Per centage	58	17	43	69	13	43	61	61

General results of the experiments.

From the Jäger rifles 10 rounds, or altogether 100 shots, were fired in 18 minutes, and about 60 shots in the time (5 minutes) during which the 10 volleys were fired.

The chamber-loaded musket gave, in close line and with 20 volleys, an effective service, greater than that of the smooth-barrelled musket, as in the following table:—

TABLE V.

Distance.	Ten Volleys: the Time of Firing them not being taken into account. Breadth of 4 Files.	12 Files.	Twenty Volleys: the Time of Firing being respectively 5 and 7 Minutes. Breadth of 4 Files.	12 Files.
At 137 yards . . .	1½ time	1⅓ time	2 times	1¼ time
At 205 yards . . .	2⅗ ,,	1½ ,,	3⅗ ,,	2 ,,

By other experiments it was shown that the chamber-loading musket is not only specially suited for independent firing, but also for firing in line; and the following may be given as the results of the practice on the 26th of March and 8th of April, 1845.

Further experiments made.

"1. With chamber-loaded muskets, 10 volleys were fired in five minutes; with smooth-barrelled muskets, 10 volleys in seven minutes; with Jäger rifles, on the 25th of March, 10 shots were fired in 15 minutes.

"2. The comparative speed of firing may be thus shown:—

	Shot.
Jäger rifle	1
Smooth-barrelled musket . . .	$1\frac{9}{10}$th
Chamber-loaded musket . . .	$3\frac{1}{10}$th

"3. Finally, referring to the tables, and taking all the distances into account, the effective practice against a target 8 feet wide is, with the chamber-loaded musket, at least three times that of the Jäger rifle, and at least three and a half times that of the smooth-barrelled musket, at the distance of 137 and 205 yards, to which that arm should be restricted."*

Precision of the Norwegian rifle.

After stating at some length the escape of the gases and other disadvantages, with their replies, as well as some other minor objections, which were answered with equal force, the Commissioners, having briefly examined the experiments made to test both the utility and the durability of this new model of a chamber-loaded musket, con-

Objections raised to a breech-loading piece,

* Royal Engineers' Corps Papers, 1849–50, p. 374.

clude by stating, that although the prejudice was at first strong against it, both soldiers and officers soon became convinced of its efficiency, and as its durability appears also fully established, they express a hope that it will be received in the army as an efficient weapon for the infantry.*

but pronounced to be an efficient weapon.

The cylindro-conical ball of the French.

The French, whose attempts had preceded those of the Norwegians, did not fail to continue their experiments, and the cylindro-conical projectile, which they used instead of the ordinary ball, possesses, as Paixhans observes, the advantage of encountering less resistance with an equal mass; consequently any piece in which it may be used, whether a musket or a great gun, will produce a shock equal to one of a considerably larger calibre, but having a spherical projectile. This fact had been ascertained by Monsieur Caron, an officer of artillery, at Charleville, in 1833; and a hollow introduced into the larger extremity of the ball, as proposed by Captain Blois, by carrying the centre of gravity farther forward, was found to give an increased range, the accuracy of which was at the same time much improved by a very simple contrivance. Monsieur Tamisier having cut channels (des cannelures, see *g*, fig. F) in the after part of the cylinder, it was found that these, like the tail of a rocket, the feathers of an arrow, or the shaft of a javelin, by opposing resistance

Channels to make their flight more accurate.

* Royal Engineers' Corps Papers, 1849–50, p. 377.

perpendicularly to the line of flight, prevented the deviation of the ball, and even caused it to resume the true direction in case of any momentary divergency.* Finally, by means of a very simple process, "the origin and peculiarity of which," says Paixhans, "are unknown to me," Monsieur Minié succeeded in causing the ball to enter the musket freely, and yet to fill up the grooves of the rifle completely, by expansion, as it passed through the bore.† The method by which these objects were accomplished has become known to us through some experiments made in the East, about the commencement of the present year. According to the "Ceylon Times," the Comte de Belloy and his friends used on this occasion two French rifles having four grooves, taking one whole turn in two metres, or 192 degrees in the length of the barrel, which is 42 inches. The ball used was of lead, 0·672 inch in diameter, 1·158 inch in height, weighing 730 grains; and with a charge of only nine grains, it penetrated and passed beyond an inch plank at the distance of 900 yards. Figure F, and the following description, will explain the nature of this projectile, which is rather smaller than the bore. The ball, *h*, consists of a cylinder having three channels, *g*,

Expansion of the balls.

Size and description of the ball.

* Constitution Militaire de la France, par H. J. Paixhans, Ancien Général de Division d'Artillerie. J. Dumaine, Paris, 1849, pp. 225, 226.

† Ibid., p. 224.

cut round the surface near the extremity, the other end of the missile being like a fir-cone. A cylin-

Fig. F.

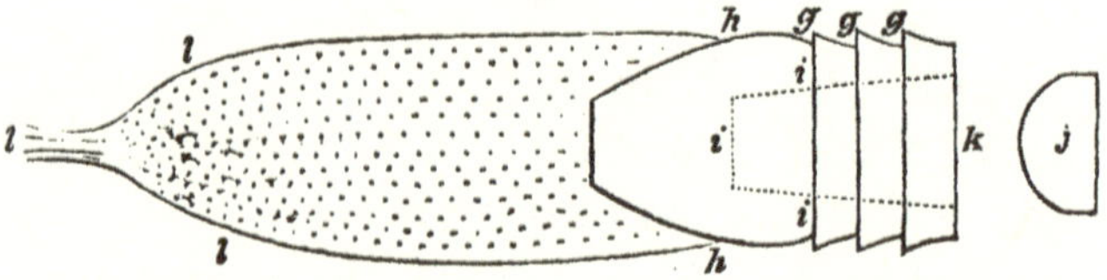

drical hollow orifice is cut into the centre of the ball, which extends, as shown by the dotted lines, *i*, from its base almost to its apex. Before placing the ball in the piece, a small capsule or thimble of sheet-iron, *j*, is placed in the aperture level with the base of the ball as at *k*, and paper being rolled over it, this end of the cartridge, with the ball in it, is dipped in grease about half an inch. When loading, the soldier bites off the end of the cartridge, shakes the powder into the barrel, reverses the cartridge, *l*, and puts the ball with the thimble end downwards into the muzzle as far as the upper channel; tears off the paper, throws it away, and then rams the ball (with the greasy part of the paper on it and the iron thimble inside) down on the powder, which is as easily done as with the common musket. In firing, the explosion, as a matter of course, forces the iron thimble up into the conical hollow in the ball, before the *inertia* of the ball itself has been overcome, and thus, by increasing its diameter,

Manner of loading the piece.

Cause of the ball's expansion

forces the lead into the grooves of the bore so completely, that the whole base of the bullet is exposed to the action of the powder without allowing the slightest windage, or any diminution of the explosive force of the powder, by which so much of the impetus is lost in common rifles.

In the piece, and its advantages.

Paixhans, in his "Constitution Militaire de la France," gives the following as the result of extensive experiments with the new rifled carbine, which only requires 4½ grains, instead of 9, of powder to propel a ball nearly double the weight formerly used.

Result of experiments made with the

At a distance of 218$\frac{8}{10}$ yards, it was found that a target of rather more than two yards square was struck 100 times in succession with the new musket, and only 44 times by the old weapon, out of the same number of shots.*

Again, at 655$\frac{8}{10}$ yards, which the common musket did not reach, the same target was struck 25 out of 100 shots by the new musket, whilst a field-piece firing the same number only struck it six times.†

French carbine at various distances.

And at 1,093 yards, when a field-piece usually diverged six or eight yards from the target, the new musket struck it six times out of 100 shots; and even at this enormous distance, it was found in the case of an experienced marksman that three

* At 200 metres. Constitution Militaire de la France, par H. J. Paixhans, Ancien Général de Division d'Artillerie, p. 40.

† At 600 metres. Ibid.

of his shots out of four took effect on a moderate-sized target; so that in this case art did more than nature, for at 1,000 yards none but a good sight could distinguish the object which the musket hit so accurately.*

Eight battalions armed with this piece.

Another French officer, on a recent occasion, writes thus:—"Nous avons en ce moment huit bataillons armées de fusil rayé à balle à culôt. Les resultats sont toujours très satisfaisants." The Belgian government, departing from the principle of the French as to loading at the muzzle, introduced a musket somewhat similar to the Norwegian, and another has been invented elsewhere which appears to combine speed in loading with the longest range and the greatest precision hitherto attained.

Improvement on the Norwegian musket.

The common musket, as is well known, has more power than is available, owing to the difficulty in striking an object; this difficulty, however, is greatly if not entirely overcome both with the French and the new Prussian musket.

The needle-igniting musket of Prussia.

The progress of the Zündnadelgewehr, or needle-igniting musket, was, however, slow at first; but the fusileers having been so armed, its adoption gradually became general, and it will probably be used ere long throughout the Prussian army. It combines the use of percussion with that of a particular kind of ball, which being conical at the

* At 1,000 yards. Constitution Militaire de la France, par H. J. Paixhans, Ancien Général de Division d'Artillerie, p. 40.

point, cylindrical in the centre, and round at the larger end, is, as in the case of the French projectile, a good deal heavier than a sphere of the same calibre. It becomes rifled as it passes through the barrel, and is propelled with much greater force than the ordinary rifle-ball, owing to two causes, viz., a suitable centre of gravity, and the more perfect ignition of the powder, which takes place in front, instead of being as formerly at the other end of the charge. This advantage, one of the greatest belonging to the change, is accomplished by means of a metal needle and a spiral spring. The spring serves the purpose of a lock, and by forcing the needle through the charge, the fulminating powder explodes it in a way which will be better understood from the following details and plate. (See page 273.)

Its peculiar ball, and means by which it is discharged.

The barrel of the zündnadelgewehr is 34 inches long, and is rifled with four grooves, taking 1¼ turn in the length, and has a high back sight A, fig. 3; it is screwed into the end of a strong open guider or socket B; the chamber properly so called is bored out in a slight degree conically from behind, C, so that when the cartridge is placed in it, the shoulder of the ball (which is of a particular shape) shall meet, and be stopped by the projections of the ribs of the rifling, the body of the ball being of sufficient diameter to fill the full depth of the grooves. Inside the guider slides an

Barrel 34 inches, with four grooves.

Description of the breech, and the needle conductor.

iron tube E, with a strong helve or handle attached, and having a space at the front end next the barrel of about 1½ inches in length, F; in the middle of this space is the needle conductor G, which is pierced with a small hole in its entire length, through which passes the needle that is to ignite the charge. This needle conductor is screwed from behind into a solid plate of iron left in the tube H; and this plate it is which (like the breechpin-piece of the ordinary musket) receives the whole reactionary force of the charge. Behind this plate, again, there is a second tube of iron, I, having a spring with double catch attached, and carrying within it an inner small tube J, which has two projecting rings on one moiety of its length, and a spiral spring on the other, K; and through this tube passes the *needle*, which is a thin steel wire pointed at the end destined to ignite the charge, the other end being screwed into a brass head, which again screws into the interior tube that carries the spiral spring. The

The trigger acts upon the needle by means of a spring.

trigger L, is of peculiar form, with a straight spring M, having two knuckle movements acting upon a ball; the first movement fires the gun, and the second admits of the whole mechanism being taken out behind, when the parts can be taken to pieces, cleaned, and put together again by a soldier in two minutes, there being no pins whatever, and no screw, except that by which the

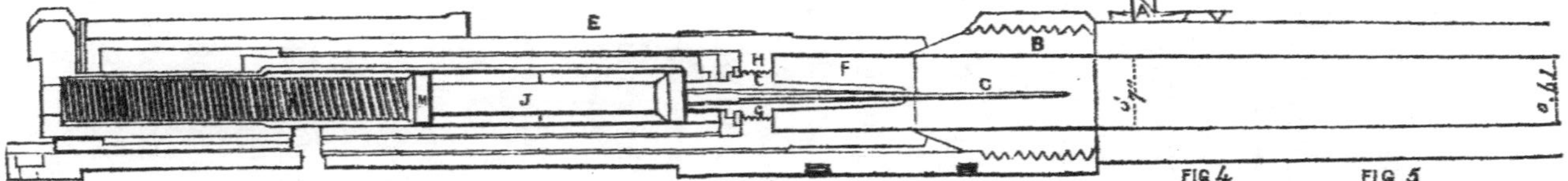

FIG 3

SCALE OF INCHES

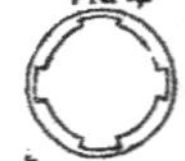

FIG 4

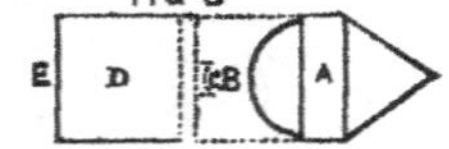

FIG 5

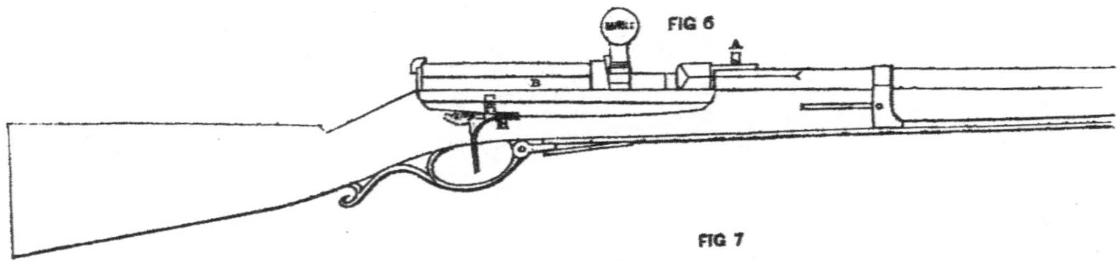

FIG 6

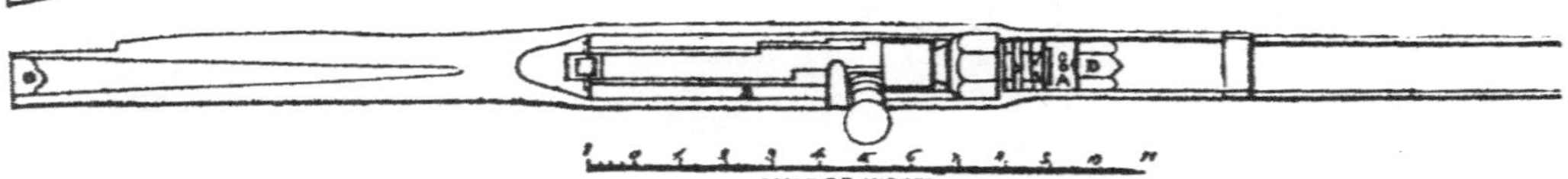

FIG 7

SCALE OF INCHES

T

needle is connected with the inner tube, and this is never disturbed, except when the needle has to be replaced by a new one.

Description of the cartridge and

The cartridge, Fig. 5, is made of one thickness, of thin but strong paper. A is the ball; B the paper bottom, with C the indentation in its lower end for the priming composition; D is the powder. The end of the cartridge at E is formed also of a single thickness of paper; through this the priming needle is forced by the spiral spring. The needle passes through the whole length of the charge of powder, and penetrates the primer C, which it ignites, and consequently the charge is lighted in front instead of the other extremity, as usual; and behind the charge there is an empty space in the sliding tube of 1½ inches long. To these two circumstances the Prussians attribute the additional range and the slightness of the recoil.

action of the needle.

Range of the piece from 800 to 1,200 yards.

Besides celerity in firing, which without over-exertion extends to about six rounds in a minute, and entire freedom from windage, by which a range of 800, or, according to some, even 1,200 yards, is obtained, there are several advantages attending the use of this weapon.

Advantages of the ball, and facilities of loading.

As already mentioned, a ball, for the same bore, is much larger than that of an ordinary musket, and being formed by pressure, it is more solid, and has, at the same time, a more correct position of the centre of gravity. Having the advantage of being rifled also, it is truer in its flight than the round bullet, especially as the powder is not crushed, as is frequently the case in ramming down an ordinary musket or rifle. Added to these advantages, it receives a greater impulse; and the pasteboard wadding, which is a part of the cartridge, assists in clearing the barrel from the effects of the previous discharge; and as the soldier can load almost as easily in a recumbent as in an upright position, he need not, when once behind cover, allow any part of his body to be exposed to the enemy's fire. In addition to the preceding considerations, the recoil is less violent, and owing to the simpler and more delicate motion of the trigger, there is much less to prevent a correct aim, so that a very accurate fire is the consequence.

Objections raised to breech-loading arms.

The objections which have hitherto been imagined are—the liability of the spring to get out of order, the divergence to the right or left to which the steel needle may be liable in passing through the powder, and the probability of missing fire when the needle gets dirty; likewise the escape of gas through the apertures, after firing has been continued for any length of time, and finally, the wear and tear of the barrel from the smoke and burnt powder issuing through the apertures at the place of junction of the cylinder with the barrel.

That some imperfections should exist may be expected as inseparable from the works of man, but they should in this case be considered in comparison with the advantages and possible effect of such an instrument on modern warfare.

These objections can be remedied in part.

The diminished power of the spring by constant use, and the divergency which may be caused to the needle, are serious, but it is hoped not irremediable evils, since both spring and needle may be renewed at a trifling expense. By having a few spare needles and springs, as one of each for eight or ten muskets, or in any other proportion that may ultimately appear desirable, the defects in question would probably be remedied, and efficiency secured; for the liability of the piece to miss fire, and the more serious defect of the escape of the gas, only take place (extensively, at least, in the latter case) after some fifty or eighty discharges, so that a general action might be fought before the piece even requires to be cleaned. It is true that the gas escaped with sufficient force to remove a trifling weight placed on the aperture, but this should not be a fatal objection to an instrument of undoubted power and precision of range. Even from a piece with a flint lock, the escape through the vent is considerable, and at any rate the evil may be lessened if not entirely removed; for since American and other pieces have close-fitting breeches, as was shown lately in the great Exhi-

Others will probably be so eventually.

bition, it cannot be doubted that the skill of our workmen will overcome the difficulty in the case of the Prussian musket.

The breech-loading musket has been partially used, and it is understood with good effect, during the late Hungarian war, and still more decisively in the north of Germany.

In one part of the hard-fought battle of Ilstedt, the Danes found themselves opposed by skirmishers armed with the new Prussian musket.

Use of the Prussian musket at the battle of Ilstedt.

"The enemy," says the Danish Commander-in-Chief, Krogh, "under cover of a bridge, fired with pointed balls (Spitzkugeln), at a distance of 100 and 150 yards. It was in vain that a couple of guns threw shells at a short range among the skirmishers; it was in vain that a body of cavalry made their several attacks; it was in vain that the endeavour was made to bring up the infantry from Oberstolk, which was now in flames, while a fierce engagement was going on in it from the house-windows and the streets. In less than an hour we suffered a great loss. The brave General Schleppegrell fell mortally wounded during the attacks; the chief of his staff, Lieut.-Colonel Bulow, was severely wounded; the commander of the battery, Colonel Baggeilsen, was made prisoner, and two of his guns taken by the enemy. Several other officers were also killed, among them Lieut. Carstensea, whilst endeavouring to rescue Captain Baggensen, and about 70 subalterns and privates; at least 90 horses were killed or taken."

Deadly effects of this weapon.

The efficiency of this weapon is now, however, being put to the test by a Committee appointed by the Commander-in-Chief, by whom the French and a variety of other muskets are being carefully examined. Amongst the number, the patent needle-gun of Sears, and the rifle invented by Mr. Lancaster, may be mentioned. The former loads at the breech, and partly resembles the Prussian musket, but has in addition a receptacle containing fifty detonating caps, which by a simple operation are brought forward successively to ignite so many charges.* The following brief description will give some idea of the construction of the latter weapon, which is simpler than the Prussian musket, though giving, it is said, an equal range.

Sears' patent gun and Lancaster's rifle.

Fig. 1. A.

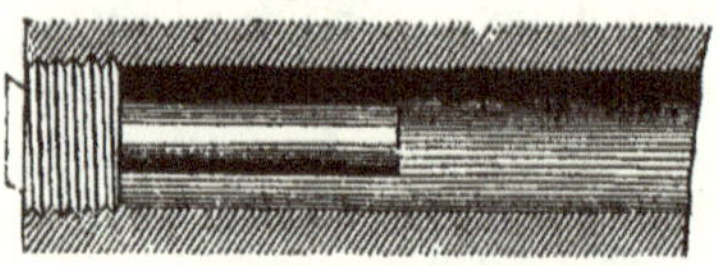

Fig. 1.

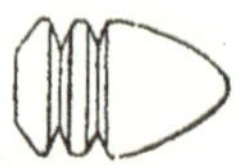

Fig. 2. A.

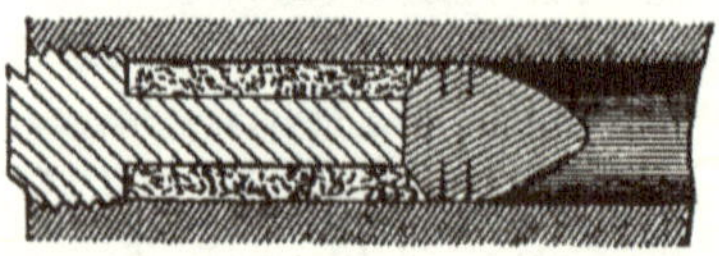

Fig. 2.

Description of Lancaster's rifle ball, &c.

Figure 1 represents the ball before it is put into the piece. The rings which will be perceived round the lower part, permit the compression of the ball, which, on being forced down by the ramrod, assumes more completely the form of the inside of the barrel.

* See Pamphlet describing Sears's Patent Needle Gun, for loading at the breech. M. A. Sears, 36, Burton-crescent, London.

Figure 1 A shows the breech end of the barrel, with the metal pin forming part of it.

Figure 2 shows the shape of the ball when its rings are compressed by being rammed home, so as to form a solid ball.

Figure 2 A shows the position of the ball prior to its being compressed by the motion of the ram-rod, and with the powder lying round the pin on which the former rests.

The new musket in one or other form will doubtless be adopted.

As the new musket, whether loading at the breech or at the muzzle, gives a more distant and a more accurate fire than is ever attained even by our best rifles, it can scarcely be doubted that in one, if not in both, of these two forms, the new weapon will be adopted in the British army, but whether of the English or of the smaller calibre of the French, requires much serious consideration.* There does not seem to be any doubt that an extreme range, with great power, belongs to both; and the weight saved to the soldier by sixty rounds of light balls is an object of paramount importance. As much more depends on rapidity of movements than on carrying a quantity of ammunition into action, the consumption of the great battles fought during the last war would be a safe guide. It is understood that the number of rounds fired has varied from

* Since this was written, it is understood that the Government has decided on the adoption of a musket on the French pattern for a considerable portion of the army.

three to about twelve. In the three days ending with Waterloo, the number of rounds fired amounted to 987,000, which, for the number of men under arms, would be from 10 to 12 each: 30 rounds, therefore, would appear to be ample for the soldier to carry, and 20 additional rounds, on an average, might accompany the army in light waggons.

The soldier to carry only 30 rounds.

Supposed effect of the new musket.

Besides a more distant execution and other advantages claimed for the new fire-arm, especially for the Prussian pattern, its advocates do not hesitate to affirm that its fire will be more formidable than that of grape-shot; that the gunners would be picked off at such a distance as to make it impossible for them to serve the guns in face of light infantry, and that it will, in consequence, supersede the use of light artillery. It is also alleged that personal conflicts, such as line against line, or column against column, will cease altogether, and future battles be decided by the effects of a rapid and destructive fire, in the precision of which, rather than on personal contact and extensive combinations, the result will depend.

The new musket favours

Since a single man can now be struck down by a musket-ball at a considerable distance, it follows that the means of defending field-works, a river, a defile, or, in fact, any strong post where the defenders can remain under cover, whilst the attacking force is exposed, will be greatly increased. In such cases, more particularly in that of a fortress, the defence will probably become superior to the attack; at least, after such modifications in the construction of fortresses shall have taken place as will give longer lines of defence, protected by a loop-holed musketry fire from those parts of the works which, in this respect, have been hitherto rather neglected.

Defensive positions.

In reference to this subject, the following remarks have been furnished to the author by Lieutenant-Colonel Portlock, R.E.:—

"Whenever two weapons of unequal powers are used in combination for the defence of the same object or point, it is evident that the range or sphere of action of the one possessed of the greater power, must be reduced to the limits within which the other can act, as without such limitation the weapon of least power would cease to have any effect, and become, in consequence, useless and superfluent. This is more especially the case with projectile weapons; and, in consequence, the length of the line of defence in modern fortresses, has been regulated on the range of the musket. Before the invention, or rather the application of gunpowder as a projective force, the inconvenience of combining together, in the defence of the same point, weapons of different powers, was scarcely felt, as there was then comparatively a very small difference in the actual ranges of the projectiles used; but when the cannon and the musket took the places of the bow and javelin the difference became very great, and it was found necessary to provide for it in practice.

Musketry regulates the

"There can be little doubt that the necessity of effecting this adjustment between the two principal projectiles used in defence, led the early Italian engineers, who (not the French) were the authors of the bastion system, to adopt that form of defensive works. By retiring the curtain they placed it in tolerable security from attack by cannon, and by making the angle of the flank obtuse, they threw the fire of the cannon of the flanks on the ditch in front of the bastinet. The defence of the curtain itself is almost entirely dependent on the musket, and as it fires over the parapet, it is evident that, in order to make that fire effective, the relief of the flanks must be regulated by the length of the curtain. Hence, the length of the line of defence and that of the front of fortification, is limited by the range of the musket which is intended to co-operate with the guns of the flanks in the defence of the salient of the bastion, and the relief of the works by the necessity of covering the half curtain by the musketry fire of the opposite flank. A neglect of these principles has frequently produced high flanks, and short imperfectly defended curtains, and it is evident that the adoption of a musket, possessed of high power and range, will at once free even the bastion system from much restriction both in the length of lines of defence and in the relief, and permit a much more efficient combination of the musket and cannon in defence. For example, if the effective range of the musket were increased from 240 to 480 yards, the length of the curtain might be doubled, and the relief of the works increased, whilst the guns of the flanks could be brought to share in the defence of the long curtain much more effectively than they could in that of a short one. In addition to these advantages, two points might be thrown into one, and one salient got rid of. These advantages would increase with a still further augmentation of range; but there is doubtless a point beyond which it could not be desirable to extend lines of defence, as sufficient certainty of aim could not be secured in quick firing. That limit will probably be about 600 yards. These remarks have been principally applied to the bastion form of fortification; but, in modern times, the tendency has rather been to adopt other forms, such as the system of tenailles,

length of the curtain.

The present lines of defence

and in these the advantage of an increased range in the musket will be still more striking. To obviate the inconvenience consequent on the short range of the musket, as well as that due to the elevated position from which it was obliged to fire, caponnières and reverse galleries were adopted, which at once brought the fire to a lower level, and shared the space to be defended between several musketry batteries or defences. The use of a musket, with increased range, will augment the advantages to be derived from caponnière defences, as it will be possible to confide entirely to them the defence of the ditches, and by the adoption of long lines to combine the gun and musket in the same caponnière, allowing to each a range of effective fire, or rather, as it should be expressed, without crippling the action of one, by forcing it to conform to the lesser range of the other; as must be the case when it is attempted to combine in fortification the gun and ordinary musket for the common defence of the same point. The difficulty of attack will be increased materially by the increased range of the musket, and especially so in the more modern traces of fortification, as that weapon will be always ready to act even when the guns have been dismounted, and it will be no easy matter to secure the breaching batteries from a reverse fire of musketry, and to keep up the communications. It would require too much space to point out the various modifications which an engineer would naturally adopt in his works, as consequent on the altered range of the musket, but they may be all referred to some one of the following advantages he gains by it:—

may be greatly extended.

Increased use of musketry.

" 1. The power of using larger lines of defence.

" 2. The diminution in the number of salients.

" 3. The power of uniting naturally strong and salient points, by simple lines, without intervening salients.

" 4. An effective co-operation of cannon and musketry in defence."

Changes in permanent works.

Beyond the preceding alterations in permanent works, and some modifications of the various arms used in the field, particularly of the artillery, by the substitution of larger guns for those of smaller calibre, it is unlikely that the science of war will experience any very great change in consequence of the use of more efficient firearms. But in closing this chapter, it should be observed, in connexion with the subject of the new musket, that if the present-sized bore should be retained either for the Prussian or the Minié weapon, there will be a serious increase of weight to the soldier; and this being the unavoidable consequence, it is worthy of consideration whether a smaller bore, and of course a lighter musket, should not be adopted.

The weight of the new musket

Undoubtedly, the Minié rifle, for instance, has a very extended range and great precision with its expansive ball, which, be it observed, owing to its elongated form, weighs an ounce avoirdupois, notwithstanding the smaller bore of the piece. If, therefore, the latter be very efficient, there does not seem to be any reason why the British soldier should be burthened with the increased weight that will be the consequence of maintaining the existing bore.

should if possible be lessened.

CHAPTER X.

ON TACTICS IN CONNEXION WITH THE NEW MUSKET.

Formations into two instead of three Ranks, proposed by Captain Wittich.—Change of Tactics to be expected in consequence of the new Musket.—Power of this arm.—Superiority of the new over the old Musket.—The Needle-igniting Musket increases the means of Defensive Warfare, and, in a lesser degree, that of Offensive also.—Formation of Troops on the Centre becomes advisable.—Keeping up a Fire with the new Musket, by alternate portions of attacking Troops.—Actions will take place at a greater distance than heretofore.—Halting whilst advancing to the Attack, and considerations connected therewith.—Position to be occupied when a part of the Army uses the new Musket.—Employment of this weapon with the Reserve.—Power of Artillery, and its superiority over the old Musket.—Range of the new Musket greater, and Fire more formidable, than Case-shot.—Use of the new Fire-arm either in Supporting or Attacking Artillery.—Influence of this weapon upon the Organization of Artillery.—Probable diminution of Light Artillery.—Greater necessity, in consequence, of Cavalry to cover the Horse Artillery.—Large Bodies of Cavalry will be less employed than heretofore against Cavalry—Part of the Cavalry to be Organized and to serve as Mounted Infantry.—Employment of this description of Troops.—Superiority of Mounted Infantry over Horse Artillery.—This new force not to be considered as Dragoons, but simply as Infantry, using Horses for locomotion.—Future Battles will require fewer Cavalry Operations.—Necessity of anticipating considerable changes in Warfare.

THE great power belonging to the new fire-arm has given rise to many discussions, particularly in Prussia and France, respecting their introduction into the armies of Europe.

With reference to the difficulty, or rather the presumed impossibility, of deploying troops under such an accumulated and destructive fire

Changes proposed by Captain Wittich.

as may be delivered by this arm, a military writer of the former country has lately proposed to change the formation of the Prussian army from three to two ranks, and also to discontinue entirely the use of heavy columns. The deepest formation he proposes to retain would be to double any portion of the line that might be threatened, so as to have four ranks to resist a charge of cavalry;* thus advocating an entire change in the tactics of revolutionary France, by adopting an extended formation, in order to accumulate a defensive fire.

The substance of Captain Wittich's description of the effects of the Prussian musket has been translated,† and is given here in order to explain his ideas of the application of the new arm more fully, especially as regards artillery.

Regarding the "zündnadelgewehr," or needle-igniting musket, Captain Wittich observes:—

"The advantages of this musket, the *general* adoption of which must necessarily occasion a complete change in the *whole tactics of war*, may be comprised under the following heads: an effective range at 600 paces, and even a considerable degree of precision at the distance of 800 paces, and with the capability of firing seven or eight shots in a minute.

An effective range of from 600 to 800 paces.

"The artillery and cavalry may endeavour to console themselves respecting the superiority of the weapon over their

* Das Fähnlein oder die Compagnie als die wahre tactische Einheit, &c., Von Wittich Hauptmann und Compagnie Chef im Königlich Preussischen 17 Infanterie Regiment, s. 16–63. Wesel, 1849.

† Ibid., pp. 74–80.

arms by the consideration that even the most perfect musket requires *marksmen*, and, therefore, that all may proceed in the old course, notwithstanding the *invention* of the needle-igniting musket; but even if practice were carried no further than it has hitherto been, the execution with the new weapons would be so superior to that with the old, as to insure by its important influence a change of tactics. Even a very inferior marksman finds it easier to hit with the new musket, on account of the delicate movement of the trigger; and it will also be found that if the *proportion* of combatants to the number of misses be taken into consideration, without allowing anything in favour of the former, the power of loading nearly six times for once with the old weapon will give a *sevenfold number* of combatants, which, independently of a greater range, is of itself sufficient to occasion a change in tactics.

Advantages of the new musket.

"The objection that the infantry cannot carry a sufficient supply of ammunition in their pouches to meet such rapid expenditure, cannot have any weight with respect to artillery and cavalry; since, when opposed to these arms, a saving of ammunition, or the means of carrying it, could scarcely be made a consideration. But in the fire of infantry against infantry this circumstance might so far be taken into account that, notwithstanding the greater range of the musket, and its efficiency at a distance, as a rule, only light infantry might be allowed to fire at the longest ranges of the weapon.

Great consumption of ammunition, particularly by the light infantry.

"In considering the details, it will be found—

"1. That the needle-igniting muskets not only give a superiority to defensive over offensive warfare not hitherto known, *but they also give increased powers to the latter*, although not in the same proportion, since, owing to the increased facility and rapidity in firing, the movements of bodies of troops would be attended with a much more destructive fire than formerly, since the light-infantry man, for instance, can direct a fire upon the enemy seven or eight

Increased proportion and proposed tactics of the light infantry.

times more powerful than at present. It becomes, therefore, necessary to consider the best organization for the *increased force* of *light infantry* which will henceforth be required to accompany advancing bodies of troops. In this respect we hold to the formation recommended in section 10* as the most suitable, *i. e.*, the companies are to be formed on the centre, and in small columns four ranks deep, leaving the necessary space between each to deploy, so that *a company* thus formed can send platoons or half a platoon into the intervals on each side, which immediately before the attack can be drawn in, and resume their places. When light infantry, as usual hitherto, are thrown out in front, the retreat of these skirmishers is greatly facilitated by the proposed intervening spaces, in which these marksmen may take post, and assist in keeping up a fire to the last moment against the attacking enemy.

Alternate movements with firing.

"A third mode of firing arising from the formation recommended in section 10,† appears to be very suitable, under certain circumstances. The odd and even numbers of the company halt alternately, and fire volleys; the two front ranks kneeling, and *two ranks* firing together alternately. Afterwards, in double-quick time, they rejoin the other companies, which during this short interval would have gained but little ground, since each volley would only consume a few seconds. As only small and simply-organized bodies would be employed in these rapid movements, they would speedily regain their positions with regard to the main body of troops; and when this shall have taken place, the other, or alternate companies, should go through the same evolution.

"We must not, however, conceal from ourselves that this mode of attack will make great demands on the physical powers and steadiness of the soldier, and therefore should only

---

* Das Fähnlein, &c., s. 46.

† Small columns, four ranks deep. Das Fähnlein, &c., s. 46.

be resorted to under favourable circumstances. It appears, however, certain that this mode of attack, by alternate discharges of musketry, must cease so soon as the enemy shall be provided with an equally-effective weapon, and that the importance of each moment that might be lost while exposed to so destructive a fire must of itself produce a new system of tactics. Part of the fire of the first discharge, for instance, is lost by the movements of the battalions when forming in columns: again, when two lines fire and halt alternately, the fire of the second is also lost, or rather delayed, till it passes the other line, and halts to fire; the comparatively small number of skirmishers that can be thrown out will be but a poor protection from the confusion which is likely to arise during such complicated movements. Moreover, the time necessary for deploying a battalion, which will be considerable even when it is formed on the centre, must be lost.

Partial use of the new musket.

Deploying under fire.

"We must, therefore, suppose that the second discharge shall take place at the moment when the first discharge has produced such an effect that the troops may advance without delay to an attack with the bayonet. This, however, is in contradiction to the acknowledged superiority which a troop armed with the new musket has in defensive over offensive warfare, particularly if the former has the slightest advantage of ground on its side. With the greater range of the new musket, the distance of all combats will also be increased: too much time would be lost if the attacking line delayed answering the enemy's fire until the second columns had traversed the distance intervening between the lines, in order to follow up the attacking movement. It would, however, be just as rational for a single attacking division to halt, in order to engage in a musket fire with the enemy before he has faced about, as it would be injudicious to engage in alternate discharges, since the latter would occasion a serious delay in the attacking movement.

Actions will commence at a greater distance, and movements become more difficult.

"2nd. Let us consider what should be the position of those

Disposition of the troops when partially armed with the new musket.

troops which are armed with the needle-igniting musket, supposing only a portion of the infantry to be thus equipped. We should reply that they might be *both in the advance and also in the reserve.* In the former, because, with the same number of troops, an equal, and, with a larger force, a much greater effect would be produced than with ordinary muskets; consequently, the troops would either be spared for the great attack, or the enemy would be driven to make a disproportionate development of his forces; and, in the latter case, he must either trust the final decision of the combat to his best troops, or give support with the latter to any particular point during the battle. And, it may be added, the new muskets belong especially to the reserve when this force is destined to act on the defensive, to take up a position in the rear of the battle, under cover of which the other troops may rally or reform, in order either to bring the combat to a different termination or to withdraw from the fight.

Heavy artillery used in preference to light.

"With respect to the 3rd consideration, the greater destructive power of heavy artillery, and the distance at which this arm is used, are the fundamental principles to which it owes its origin; and these principles have been carried out at the expense of hand-grenades as well as of other fire-arms. Hitherto, case-shot has compensated for the small number of guns, since the loss in point of range was made up by the number of shots it discharged at once; maintaining as it did, at the same time, an effective range then much surpassing that of infantry fire.

Musketry superior to case-shot.

"But the range of the new weapon not only exceeds that of case-shot, but also, owing to the increased numbers to be used, it will have such perfection and efficiency that it will have quite the effect of case-shot. Moreover, the increased firing that will be the consequence of the rapid loading of the new musket, compared with the number and effect of case-shot (even admitting that a greater destructive power may still be conceded to the latter), it may still be proved that artillery with case-shot must give way to infantry armed with needle-igniting muskets. But so long as one of the hostile parties is provided with this superior weapon, it will naturally contribute greatly to the support of its own artillery; since, under such a cover, the latter will gain a superiority over the enemy's cavalry and infantry. We are thus shown a third mode of making use of the new musket, which will be practicable even if only one portion of the infantry were to be thus armed; and the united power of the two arms must have such a decisive effect that the advantages of the musket cannot be safely dispensed with.

Field guns will not be able to resist

"Should each of the contending parties be in possession of this superior weapon, even though only a portion of each force be armed with it, its partial use even must demonstrate the advantages that may be obtained by its means.

"Let us further consider that artillery is only used against bodies of infantry, while the light infantry can take up positions from 300 to 400 paces in advance of the main body, unless they are prevented by the approach of the enemy's cavalry; added to which, the fire from two platoons of light infantry, at 600 paces, will be so effective, that a hostile battery, without similar protection, would be unable to take up a position, and much less maintain it; and it is thus shown that by the use of the new musket the enemy's artillery will be kept at a distance of more than 1,000 paces from the main body of troops. Even the approach of hostile cavalry will make but little difference in such a case, if we consider that a fire can easily be opened by the light infantry, which can be rapidly formed for this purpose into small, compact bodies, under the protection of their own artillery, and in such a position that lines of infantry can fire upon the enemy's cavalry, without hurting their own people. As such a state of things must necessarily prove disadvantageous to the enemy's cavalry, without their even being able to relieve their artillery from their dangerous opponent, it is highly

the powerful fire of the new musket.

Cavalry cannot protect the guns from this formidable attack.

necessary that the light infantry should be encouraged to feel perfect confidence in their weapon, and that they should also be exercised in the rapid formation of small, compact, regular bodies. Indeed, all that we now expect from bodies of sharp-shooters may be obtained with certainty from small, but well-practised bodies of troops, rapidly formed, and using the new musket.

Systematic organization required for light infantry.

Modifications in the artillery service.

"If, therefore, the assumption be not too bold that artillery could scarcely advance within 1,000 paces of a body of infantry thus armed, it follows that the whole organization of the artillery must be influenced by the general introduction of the new musket. It is, therefore, presumed that the light artillery must be diminished, and, as the necessary consequence of coping with the new arm, *the heavy artillery proportionally increased.* Although an increase of heavy field artillery thus becomes unavoidable, some diminution of the arm at large should take place, partly on account of the expense, and partly on account of the *slowness of the movements* which the substitution of heavy for light artillery would entail on the other troops.

Effect of these changes on horse artillery.

"The horse artillery, however, might possibly retain their present guns of light calibre; but the chief use of horse artillery will be changed, since its principal object will no longer be to support cavalry, but to seize the moment when, by a rapid movement, it can not only act with great effect at short distances, but also maintain its position, even in situations of danger, to the last moment. The horse artillery and cavalry will, however, be obliged to change the parts they have hitherto acted in the field, since a greater number of cavalry will be required to cover horse artillery than has hitherto been the case. Besides this, it is evident that cavalry, as well as horse artillery, will be quite indispensable to complete the success that may be obtained in large as well as in small engagements.

"It is not without the greatest apprehension of drawing

upon ourselves the ridicule of artillery officers that we have ventured to make the preceding remarks upon their service; and it is with still greater diffidence that we now proceed to say a few words on another arm.

"As cavalry will cease to become formidable or even dangerous to infantry as soon as the latter is armed with the new musket, it follows that *a great portion of the former may be dispensed with*, since the decision of a battle will now depend more than ever on the infantry and artillery. In future, large bodies of cavalry will rarely be employed against cavalry, although we are very far from wishing to limit the operations of this arm to escorts and covering the flanks, or even to the pursuit of the enemy. It is presumed that the main body of the cavalry may be limited to a small *reserve*, and that the so-called *division cavalry* should be strengthened.

Cavalry will be inferior to infantry thus armed.

"We would, however, endeavour to render the adoption of the new musket more complete, by arming a considerable portion of the cavalry with this weapon, and converting them into *mounted infantry*, the horse being simply the means of rapid locomotion. Such a force would be of inestimable value; for instance, in the case of the advanced guard reaching a certain post before the enemy, which might be occupied with marksmen, and thus be enabled to oppose an approaching battery at a greater distance, and for a longer time, in consequence of having the power of retreating quickly. For the same reason, the artillery ought to have the protection of mounted infantry, which would give it a high degree of confidence, even when opposed by infantry armed with the new musket, more especially as infantry coverers cannot follow up the movements of the artillery when the gunners are mounted for the sake of rapidity.

Introduction of mounted infantry.

"When cavalry is engaged both with cavalry and infantry, this mounted infantry would be of the greatest service. For instance, if the cavalry, shortly before making an attack, were

Objects to be accomplished by

to throw out their coverers on both sides, so as to flank the enemy's line, the most decided effect might be expected, unless the enemy is also able to defend himself with a similar weapon. In this case, its use becomes still more indispensable, in order to paralyse the effect of that of the adversary. Should horse artillery be on one side only, and mounted infantry on the other, we consider the superiority of the latter to be undeniable. In attacks made by cavalry or infantry, the former can throw their mounted riflemen on the flanks of the enemy, with the double object of driving in his light infantry and of drawing a portion of the fire of the body attacked on themselves. Although we consider the success of cavalry against infantry to be impossible, so long as the latter are not discouraged, and use their new weapon effectually, we believe the means we have indicated to be the best calculated to facilitate success.

this new description of force.

"We do not, however, conceal from ourselves the technical difficulties which such a body as mounted infantry would have to overcome, and we would especially avoid giving the impression that we contemplate something like an hermaphrodite troop, such as dragoons are at present constituted, or even flanker-platoons. Dragoons prefer being called cavalry to infantry, and their discipline bears more relation to the former than to the latter service; and still less do flankers answer our idea of mounted infantry; for in the same way that the cavalry soldier relies, under all circumstances, upon his sabre or lance, the mounted infantry man ought always to trust to his musket. In fact, the horse is only to carry him rapidly to the particular spot where he may dismount and use his musket with most advantage, leaving for the time to the horse-holders the charge of his means of making a rapid retreat when necessary. The services of these troops must, therefore, be exclusively confined to those purposes which are incontestably necessary and may be accomplished by them, for the real object would be lost sight of if the love for cavalry,

The proposed formation is not to be in any respect like dragoons or flanking cavalry, in preference to infantry service, were infused into such a corps. It is a different thing to form a horseman who is fit for cavalry service, or to train a soldier, to whom the horse is nothing more than the means of transport.

but purely rapid-moving infantry.

"The numerous improvements in fire-arms lead us to conclude that the art of war, which since the discovery of powder has assigned the decision of battles chiefly to the infantry and artillery, will go still farther, and shake off this remnant of the ancient combats of knights; and that military tactics will eventually set aside that part of their system which the cavalry mode of combat has hitherto imposed; for, owing to the increased efficiency of the weapon to be used, combats, although carried on from a distance, will become more murderous, and will therefore be more quickly decided.

Supposed change with respect to the usual attacks of cavalry.

"The perfection of fire-arms, as well as strategic operations, by means of railroads, have much increased the value of time; and as new problems meet us everywhere, it is most necessary that we should endeavour to solve them, in order that the necessity of putting them in practice may not come too suddenly upon us, when we should have to buy our experience most dearly; for such experiments are but too often purchased by the sacrifice of human life."*

We should be prepared for these changes.

The preceding observations give succinctly, though it is hoped accurately, the ideas of Captain Wittich, in connexion with which it is proposed to make some remarks elsewhere, when noticing what has been stated by Paixhans regarding the probability of the new musket driving light artillery out of the field.

* Das Fähnlein, &c., ss. 74–80.

CHAPTER XI.

ON THE EXPECTED CHANGE IN TACTICS.

Instruction of Marksmen at Vincennes.—Danger from the new Musket to Field Artillery.—A Company of Marksmen at 650 yards might drive Light Artillery out of the Field.—The use of the new Musket will become general.—Mounted and other Light Infantry will be equally available on each side.—Probable effects of the new weapon on Warfare and Tactics generally.—Marksmen cannot maintain an isolated advanced position.—Battles must still be decided by General Attacks.—Some modification of the three Arms will be caused by the introduction of the Breech-loading Musket.—Employment and use of Mounted Infantry.—The Irregular Horse of India are an example.—A Breech-loading Carbine necessary for the Cavalry.—Superiority of Artillery.—Moral effect of Rockets and Spherical Case-shot.—Large Calibres desirable.—Guns to be more carefully protected.—Baron Wahrendorff's Breech-loading 68-pounder, and its general range.—Mr. Lancaster's Elliptical Bored Cannon.—Proposed use of Leaden Expansive Shot for great Guns similar to the Minié Rifle.—An increased Range may be expected.—A more general use of Rockets and Shells recommended.—Hale's Rockets.—Spherical Motion obtained.—Experiments in Switzerland and America.—Purchase of the Invention by the United States' Government.—Supposed improvement of the Rocket in France.—Official Statement concerning Hale's Rockets.

In the last chapter some idea has been given of the changes in tactics which the Germans expect will be the consequence of the distant range of the new musket: the present chapter will contain the views entertained by talented French officers concerning the weapon which they have adopted, and a few remarks from the author in connexion with this important subject.

The various improvements which have been effected in the construction of the French carbine-musket and its ball by the ingenuity of Delvigne, Thouvenin, Tamisier, and Minié, since 1838, have been already noticed. It is not too much to say that the range and accuracy of this piece, like those of its Prussian competitors, are so great that, unless provided with a similar weapon, the physical superiority and moral courage of an enemy, however great, would be unavailing. It enables the adversary to remain beyond the range of an ordinary musket, and to keep up a deadly fire, such as that of the chasseurs of Vincennes against the defenders of Rome. The terrible execution of these sharpshooters in picking off the Roman gunners during the late siege must not, however, be altogether attributed to the greater efficiency of the carbine, for the ordinary rifle, or even a common musket, would have caused a serious loss to an enemy, if used by soldiers who had been trained with the same care as a special arm.

Improvement of the French carbine.

The chasseurs of Vincennes, and

The means of imparting instruction adopted at Vincennes claim a moment's attention, in consequence of the manifest advantages of practising at unknown ranges, instead of the usual custom, whether with artillery, cavalry, or infantry, of firing at specified distances. At Vincennes, the soldier, after attaining some proficiency according to the old routine, is practised in estimating distances as if before an enemy, in the following

the method of their instruction.

manner; and in order that the knowledge imparted at Vincennes may be extended to the whole army, at least one *sous officier* is brought there from each regiment.

Calculation of distances.

The distance of an object is ascertained by a simple triangular instrument, having a scale, the basis of which is the ordinary height of a man at a certain distance, and as this distance is augmented or diminished, the scale thus obtained will cover more or less than the supposed height of a man. Without going into details, it will be sufficient to say that such means are more accurate, but less useful, than measurement by the eye, based, as the latter should be, as in the case of a sportsman, on practice.

Pacing distances.

The great desideratum in firing must be the means of verifying the distances estimated, which is done either by a measuring line, or by pacing from point to point. A squad, usually of sixteen men, is taken to the ground by an under-officer or corporal; and the latter having measured and marked distances of 50, 100, 150, and 200 metres, &c., the men are practised in moving from one spot to another, and the number of their paces being ascertained between certain points, the instructor causes them to lengthen or shorten their paces, so as to go over a given distance in a certain number of paces. When the soldier has ascertained how many paces are required for a certain fixed distance, he is made to advance at the same pace, and call out the distance when he arrives at 50, 100 metres, and so forth, till it becomes easy to him to estimate distances with sufficient accuracy to fire at each with precision.

Distances determined by the eye.

The mode of judging distances by the eye alone is of the same nature, and as follows. The instructor having placed four of the sixteen soldiers at distances of 50, 100, 150, and 200 metres, facing the remainder, who are supposed to be the main body, the men of the latter are caused to observe such details of each man's dress as can be distinguished at the several distances respectively. Having carefully noticed the differences which exist, the instructor practises the men at certain distances which are unknown to them, in order that they may apply the knowledge they have acquired, and reduce it to practice with groups of men as well as individuals. Rules are also laid down regarding the line of sight to be used in firing, according to what part and how much of each man's dress can be distinguished.* After the soldiers have been sufficiently practised in this way, their correctness in judging of distances is subjected to another kind of test. A man runs forward, and places a target at some distance unknown to the

Practice in ascertaining distances.

* Instruction Provisoire sur le Tir à l'Usage des Bataillons des Chasseurs à Pied. Paris, J. Dumaine, 1848, pp. 60–80.

Guessing distances.

others; each man is then called upon in turn to name the distance, and the answers are recorded in a book, while much merriment is caused by those whose replies are very wide of the mark. This kind of practice takes place at all distances particularly at those between 500 and 1,000 paces, and is continued till all are moderately skilful. Firing then takes place, also at distances unknown to the men; and those who are most successful are rewarded by promotion, and become themselves the instructors of others.

The preceding description is in some degree necessary with reference to the supposed influence of the 14,000 chasseurs of Vincennes upon an enemy's artillery.

Batteries of artillery supposed to be put hors de combat by musketry fire.

The idea that light artillery will not be able to maintain its position against trained sharpshooters armed with the new musket is thus expressed by a distinguished writer:—"In penetrating a column," says Paixhans, "and overcoming substantial obstacles, or when at a considerable distance, the cannon will doubtless retain its ancient superiority; but the fire of marksmen against batteries must have the most fearful effect. At 650 yards, for instance, almost every shot will take effect on the horses and men attached to a battery, which must, in consequence, be speedily silenced: but," he adds, "it is true, that a company of expert marksmen, posted in front of our batteries, will produce the same effect."* It is stated in another place that the new musket has an equal range and greater precision than field artillery, and that a company of marksmen can produce an equal effect at less cost than a battery of artillery, which would be soon rendered quite inefficient.†

Effect of the new musket.

Giving all due weight to the arguments of the German soldier, and also to those of the well-known and highly-talented Paixhans, it is difficult to imagine that a battery of artillery could be thus summarily placed *hors de combat*; unless, indeed, it were to be left in an isolated position, without the protection, or even the ordinary support, which it should at all times receive from one or both of the other two arms.

The general use of the new musket.

Should the new musket realize the expectations even of its most moderate partizans, its use will doubtless become general throughout Europe, and it will no longer be possible for one army to throw out clouds either of mounted or ordinary light infantry, much less of single companies of these, as has been imagined by the preceding authorities, without being opposed by similar means. But even if we suppose for a moment that in some cases it could be otherwise, and that the forces receiving an attack should be unprovided with

* Constitution Militaire de la France, par H. J. Paixhans, pp. 41, 42.
† Ibid., p. 72.

Marksmen must be driven in.

light or other troops armed with this weapon, it is not to be imagined that an enemy would be permitted to retain such positions as would enable him to pick off at leisure the artillerymen serving their guns, and the officers belonging to the rest of the troops. Such an unequal contest could not be allowed to continue; for if one side had neither cavalry nor light infantry to drive in such marksmen as might be about to give this annoyance, he would still have the resource of advancing *coute qui coute* to bring on a general action.

Distant battles of musketry, without

Presuming, however, that similar offensive and defensive means would be at command on both sides, the contest in the first instance would resolve itself into one of light troops, whose attacks, being mutually supported, would (agreeably to the supposition that the new arm must supersede personal contact) be succeeded by a continuous fire from two extended hostile lines, till greater destruction on one side should lead to victory on the other.

closing, are not likely to be the result.

Tactics of this kind, with two long extended lines, are not, however, likely to follow the introduction of a more powerful engine, nor is an incessant fire of musketry more likely to become the sole means of gaining a battle in these days than it was when the greatest of all changes in warfare occurred by the use of gunpowder as a propellant. We all know that the substitution of the matchlock for the arrow did not by any means put an end to close attacks, although, comparatively, a much greater range was the consequence than that now under consideration.

Close attacks continued after the introduction of fire-arms.

This, indeed, does not appear to be sufficiently great to enable light troops to act in the manner contemplated; since, unless closely supported, they would in turn be exposed to a rapid attack of cavalry or mounted infantry. But it must not be forgotten that spherical case-shot from 9-pounders would take full effect on the enemy's musketeers at a distance beyond the range of his muskets; so that a few rounds of the former, with some rockets and rolling shot, must drive such parties in before they could take their intended position, and of course previously to the action becoming general. Except, therefore, in the supposed case of a battle to be decided entirely by musketry, an attack must, although attended with much heavier loss, be made, as heretofore, by infantry or cavalry, under the protection of a concentrated fire of artillery playing upon some part of the enemy's line. Therefore, beyond ceasing to expose dense columns, which even under ordinary circumstances have frequently failed in Spain and elsewhere, a modification of the tactics of the different arms will probably be the only changes caused by the introduction of the new musket.

Range of the new musket not equal to that of spherical case-shot.

Dense columns will be discontinued.

As concerns infantry, a greater proportion and a more general use of light troops becomes indispensable as the consequence of the new weapon, whether the breech-loading musket or its rival; more particularly by introducing that particular description of troops which, next to the artillery, received Napoleon's especial care, namely, chasseurs à cheval* Mounted infantry, when superior marksmen and thoroughly exercised, would, as already stated, pages 293 to 295, have a great influence on those operations, the success of which depends upon celerity in securing a post, or the passage of a river, &c. The important services recently rendered by the irregular horse in India may give some idea of what can be done, and the writer has seen the men of Skinner's Horse break several bottles by the fire of their matchlocks, as they passed in succession at a gallop.

Use of mounted light infantry

and of irregular cavalry.

The cavalry will necessarily receive an improved carbine, which in their case should load at the breech, and thus give a degree of facility that must, as far as fire-arms are in question, be more advantageous to this arm than the proposed change can be to the infantry. A larger proportion of skirmishers, and much practice in the evolutions connected with such duty, whether

A breech-loading carbine is desirable for the cavalry.

* A fair proportion of such a force would speedily put an end to hostilities in Southern Africa.

mounted or dismounted, would seem to be imperative, in order to cope with the increased power of the infantry; and a breech-loading carbine is necessary for this purpose, even if the weapon for the infantry is to be of the French rather than of the Prussian pattern.

Destructive and moral effects of artillery unimpaired.

If the result of experience should bear out the opinions which the writer has ventured to express, the artillery will maintain in all its force that moral effect which, irrespective of their destructive powers, is produced on warlike operations by great guns inspiring fear in the enemy, and a proportionate degree of encouragement in those who use these weapons judiciously.

Sharpshooters could not maintain their position.

It is true that Voltaire tells us* that the sound of whistling balls was music to the youthful monarch of Sweden; but the indifference to danger evinced by this celebrated warrior cannot be always expected: for although ordinary courage seldom fails, it is more than doubtful whether the best sharpshooters would maintain their posts in the face of rolling shot, the stunning whiz of a flight of ground rockets, and the bursting of spherical case-shot.

Field artillery to be of heavier calibre.

It has already been supposed that the use of the lighter pieces will be chiefly, if not entirely, confined to the horse artillery, and that field

* Histoire de Charles XII., p. 34; edition de P. N. Rabaudy. Londrès, 1803.

artillery should mostly consist of 9-pounders, with a proportion of 12-pounder howitzer batteries and some rockets, in order to produce a decided effect on the enemy when still at a distance. With reference to closer quarters and the use of the new musket, the artillery must be even more carefully covered than has hitherto been thought requisite; and possibly it may also be desirable to strengthen the gun detachments by two or three additional men, who, in case of the batteries being accidentally deprived of the usual covering party, might endeavour, as marksmen, to cope with scattered sharpshooters. This would enable the guns to continue their fire uninterruptedly, whether to resist an attack or to cover one about to be made on some important part of the enemy's position, and thus enable artillery, even without an increased power of range, to maintain in some degree its wonted superiority over musketry.

in order to produce a decided effect.

A breech-loading rifled great gun,

Reverting to the earliest kind of guns, a project for loading great guns at the breech was brought forward a few years ago, after the plan of a Piedmontese officer, M. Cavalli, with a certain degree of success, having gained an increase of one-quarter in the range; with, as a matter of course, the advantage of being loaded without exposing the men so much as at present, whether at the port-hole of a ship or the embrasure of a battery. To the plain bore used on this occasion, one that was rifled, the invention of Baron Wahrendorff, succeeded, and was applied to a 68-pounder. The ball is cylindro-conical, with projecting wings, something like the rifle ball invented by Mr. Lovell, in 1844; and being introduced at the breech, it is kept in its place by means of a transverse iron wedge. Considerable accuracy of firing appears to have been attained, with a greater range, by about 1,200 yards, than that of an ordinary gun.

introduced by Baron Wahrendorff.

Great range attained in consequence.

Another great gun is being brought forward by Mr. Lancaster, who, by changing the usual construction, with a view to increased accuracy of flight, has adopted an elliptical bore, and an elastic wrought-iron cylindrical shell, with the advantage of possessing the principles of concussion, as well as percussion, and at the same time accuracy of fire. It is understood that the merits of this invention are about to be tested by order of the Master-General of the Ordnance.

A great gun proposed by Mr. Lancaster.

The want of pliability inseparable from iron would in this respect give a manifest superiority to leaden balls, which, by adopting a modification of those about to be introduced for muskets, might not only be rifled during the discharge, but at the same time overcome the windage so completely as to obtain an extreme range with a 6-pounder, or even a smaller gun. There does not seem to be any doubt that a cylindro-conical

Supposed use of leaden balls for

ball, having, as already described, pp. 266, 267, circular channels or notches, with a hollow in the centre of appropriate size, would, as in the former case, be forced into the grooves of the bore by the action of the iron thimble, and at the same time obtain the spiral motion, which it was stated by Robins, as far back as 1742, must produce a complete change in the practice of war. Although there are serious disadvantages connected with the use of leaden balls in certain cases, there are others in which they would answer to the same extent as the rifle ball, and give a proportionably increased range; nor does it appear by any means impossible to apply the same idea even to spherical and ordinary case-shot.

great guns, and their advantages.

Increased use of shells and rockets

But whatever may be the ultimate success of these and other improvements in artillery, a more general use of rockets and shells, fired horizontally, may be confidently expected; and, as a natural consequence, their substitution, to a great extent, for solid shot.

As justly observed by Paixhans,* more attention should be given to the rocket, than has been the case hitherto, excepting in Austria, where the result of their efforts remains a profound secret.

Mr. Hale's improved rockets.

The serious objection of the tail, or stick, in the case of this weapon is understood to have been overcome by Mr. Hale's invention, who has, it

* Constitution Militaire de la France, Dumaine, p. 70.

appears, by some means (unknown to the writer) succeeded in placing inside the case, not only the materials which give impetus to the projectiles, but also, combined with the means of propulsion, the power of giving the missile a spiral motion, commencing at the instant the rocket begins to pass along the tube through which it is fired.

The federative government of Switzerland caused extensive experiments of the power and advantages of this weapon to be made, under the superintendence of a committee of artillery officers, and a number of rockets were fired from a stand at 5, 10, 15, 25, and 27 degrees of elevation. A target was placed at 1,200 paces, and the rockets used on this occasion were 10-pounders, the smallest of Mr. Hale's invention, his largest being 100-pounders. One fired at 5 degrees went on like a serpent, and never rose above 6 feet from the ground. Another, at 10 degrees, made its first graze at 500, the second at 1,300, the third at 1,900 paces, and without rising more than 9 feet from the ground during its flight. One discharged at 15 degrees first struck the ground at 1,200 paces, the second time at 2,200 paces, and when rising again the shell exploded; its greatest lateral deviation was about 50 paces.* A single 10-pounder rocket was fired at Woolwich by Mr. Hale, in the pre-

Result of experiments recently made in Switzerland.

Experiment at Woolwich, 1849.

* See the Army and Navy Register, March 1, 1849.

sence of some of his friends, on the 30th of March, 1849. A wrought-iron tube, moving on a cast-iron stand, was used on this occasion, and the rocket, being discharged at an angle of 20 degrees, without previously grazing, penetrated 10½ feet into wet, close, loamy soil, at the distance of 5,200 feet, which is scarcely less than the effect of a 12-pounder shot at the same distance.

The invention purchased by the United States.

It is understood that the government of the United States, after testing the efficiency of Hale's rockets by a series of experiments made under the direction of a committee of artillery officers, purchased the secret, and used this instrument with the greatest advantage during the late Mexican war. It is believed that the Russian government has also acquired the secret, and that the French artillery officers have recently improved the power of their rockets very considerably. The following are the official American documents regarding Hale's rockets:—

"*Washington, December* 1, 1846.

"REPORT of the joint Board of Officers of the Army and Navy, appointed by the Secretaries of the War and Navy Departments, for examining Hale's rockets.

"The Board have tried by firing on land and water 2¼-inch rockets, presented to them for that purpose by Mr. Hyde, by whom they were procured from the inventor, Mr. Hale.

"From these trials the Board have arrived at the following conclusions:—

"1. The effect of the rockets, with regard to range, force,

and accuracy, is at least equal, and probably superior, to that of the ordinary Congreve rocket of the same size.

"2. The fact of this rocket being without a stick gives it an incontestable superiority over the Congreve rocket, with respect to facility, convenience of service, and, especially, for use on board of armed vessels or boats.

"3. It is therefore recommended that an arrangement be made with the proprietor for the purchase of the full instructions requisite for making these rockets.

"4. The evidence of such information being fully and correctly communicated should be the making, under these instructions, of a certain number of these rockets, of at least two sizes (say ten each of 2-inch and 3-inch), which shall perform as well as those exhibited to the Board.

" *Officers of the Army.*

" (Signed) JOSEPH G. TOTTEN, *Colonel and Chief Engineer.*
G. TALCOTT, *Lieutenant-Colonel of Ordnance.*
A. MORDECAI, *Captain of Ordnance.*

" *Officers of the Navy.*

L. WARRINGTON, *Commodore.*
THOMAS A. P. C. JONES, *Captain.*
L. M. POWELL, *Commandant.*
A. B. FAIRFAX, *Lieutenant.*
Honourable JOHN Y. MASON, *Secretary of the Navy.*"

"SIR, *City of Washington, December* 9, 1846.

"THE joint Board of Army and Navy Officers, appointed by us to examine Hale's rockets, have made a confidential report, of which a copy is enclosed. We are authorized by the President of the United States to submit, for your acceptance, the proposition recommended by the Board in the Report.

"Will you be pleased to make known to us your determination? No delay which can be avoided will occur, if you accede to the proposition in applying the proposed tests, and your communication in writing of the necessary instructions will be treated as confidential.

"Respectfully,

"Your obedient Servants,

"(Signed) W. L. MACEY, *Secretary at War.*
J. Y. MASON, *Secretary of the Navy.*

"Mr. J. B. HYDE,
Now at Washington."

"REPORT of the trial of rockets of Hale's patent, made at Washington Arsenal, January 5, 1847.

"These rockets, presented for trial in pursuance of the former recommendation of the Board, were made at Washington Arsenal, under the direction of Mr. Hyde.

"There were fifteen 3-inch rockets, two of them with shells in the head, and thirteen 2-inch rockets, four of them with shells; of these, the following were fired in the presence of the Board:—

"Six 2-inch, with shot-heads.
Three ditto, with shells.
Four 3-inch, with shot-heads.
Two ditto, with shells.

"At various elevations from 14° to 35°.

" The accuracy of direction and the ranges were, in all respects, satisfactory, and the Board are of opinion that these rockets were quite equal to those before exhibited to them by Mr. Hyde, as having been made by Hale, the inventor; and that Mr. Hyde has, therefore, 'Fully and correctly communicated the necessary instruction for making Hale's rockets.' The trial of the rockets was witnessed by the members of the Board, whose names are signed to this report.

" *Officer of the Army.*

" (Signed) A. MORDECAI, *Captain of Ordnance.*

" *Officers of the Navy.*

L. WARRINGTON, for himself,
and T. A. P. C. JONES.
L. M. POWELL, *Commander.*
J. B. FAIRFAX, *Lieutenant.*

" *Washington, January* 6, 1847."

" AMONGST the documents printed for the use of the Senate of the United States of America, by a resolution of the 7th December, 1847, is the following short report by Captain A. Mordecai, commanding the Washington Arsenal:—

" *Mr. Hale's Rockets.*

" In the month of December last, a war rocket of a new kind, invented by Mr. Hale, of England, was offered to the notice of our Government, and a mixed Board of officers of the army and navy was appointed to test the invention.

" Experiments were accordingly made with some of Mr. Hale's own rockets, and with others made at the Arsenal according to his specification; the results of these trials were so satisfactory that, on the recommendation of the Board, the right of using the invention was purchased by the Government.

"The peculiar advantage of this new projectile is that of having its directive force in the body of the rocket, thus dispensing with the use of the cumbrous stick attached to the Congreve rocket. About 2,000 of these rockets, of the calibres of 2½-inch and 3½-inch, have been made at this Arsenal, and the trials which have taken place, from time to time, seem to confirm the favourable opinion at first formed, that, in extent of range and accuracy of direction, they are equal, and perhaps superior, to the common rockets of equal size. A report of the trial of those which have been sent into the field is looked for with interest."

The other question, the general use of the howitzer, will form the subject of the following chapter.

CHAPTER XII.

ON THE GENERAL USE OF THE HOWITZER.

Changes effected in the French Artillery Service by Vallière and Gribeauval.—Prince Louis Napoleon's proposition that a 12-pounder Howitzer shall be the only Gun used in future.—Advantages of this simplification of the arm.—Early use of the Howitzer in India, and again in Germany in the 16th century.—The Russians followed the Turks in the use of the long Howitzer.—The French adopted this piece after the Battle of Smolensko, and Paixhans' Gun followed.—Comparative Weights of the old Guns and the new Howitzer.—Advantages of the latter for general use.—Experiments with both pieces at Metz, Strasburg, Toulouse, and Vincennes.—Tables giving the results with single Guns, and with Batteries also.—Objections made to the Howitzer Gun.—Answers to these Objections.—Partial use of the Howizer in the British Service.—Proposed Extension of this kind of Arm, with a 9-pounder Howitzer also for Rapid Movements.—The British Artillery is not in proportion to the other two Arms.—Less Field Artillery employed now than formerly.—More Gunners required for Home Defence.—The Horse, Field, and Garrison Artillery Services should be considered as a whole, and be in proportion to the Line.—Field Artillery to become permanent.—Duty to be taken by Battalions under younger Colonels.—The Artillery and Engineer Services to be placed under the Commander-in-Chief.—Officers of these Services to be eligible for Staff and general Employment.—Horse Artillery to be maintained.—Captain Favé's Remarks on the Horse Artillery and Field Batteries.—Effects produced by Artillery in the Field.—No reorganization can serve the present Superior Officers.—Facility of the proposed reconstruction, and its advantages.

Modifications of the French artillery service.

We now proceed to glance at the great change in artillery which is being subjected to the test of experience in France. It was only in 1732 that the use of the 32-pounder field-piece was discontinued in that country on account of its weight; and the 24, 16, 12, 8, and 4 pounders only were retained, the last being attached to the different regiments.

But although Vallière established an uniform pattern for the guns throughout France, the carriages for pieces of the same calibre varied in the different departments, and there were as many different patterns as there were arsenals. The axletrees were of wood; and these, as well as the wheels, poles, limbers, tires, and, in short, everything except the gun itself, differed according to the locality of their construction.*

Changes introduced in 1788.

Such was the state of the artillery up to about 1788, when Gribeauval, notwithstanding great opposition, succeeded in separating the field from the siege artillery, and secured to France the benefit of the experience he had acquired during his service with the Austrian army in the seven years' war. He compared the state of the artillery in France with that arm among other nations, particularly with that of his late opponents the Prussians; and feeling the great importance of simplicity, he at length overcame all opposition, and succeeded in reducing the number of field-pieces in France to a six-inch howitzer and three kinds of guns, viz., 12, 8, and 4 pounders. These he provided with carriages suitable for each calibre; the several parts of which were so carefully

* Nouveau Système d'Artillerie de Campagne de Louis Napoleon Bonaparte, &c., par le Capitaine Favé. Dumaine, Paris, 1851, pp. 9, 10.

constructed, in accordance with a fixed pattern, that in case of accident the wheels or any other part could be replaced by others of the same kind, taken into the field for the purpose. Seven sizes of wheels and three kinds of axletrees sufficed for the field-carriages of the artillery.*

Carriages of the same pattern.

In the organization adopted in 1827 the 4-pounder was suppressed as being inefficient, and only 15 and 16 pounder howitzers were retained, with improved carriages, carrying a greater proportion of ammunition than 8-pounder and 12-pounder guns.†

The use of a howitzer of the latter calibre has been lately advocated by Prince Louis Napoleon Bonaparte, under the impression that such a piece, of a suitable construction and of sufficient strength, might fulfil every purpose connected with war in the field.

12-pounder howitzers proposed to take the place of all other guns.

This proposal opens a large question, such as the undeniable advantages of having fewer calibres as well as the kinds of ammunition, which should, however, be carefully balanced against the disadvantages, which are ever inseparable from even the most beneficial changes. It will be recollected that one of our troops of horse artillery had three kinds of guns in Spain, viz., 3-pounders, light 6-pounders, and 5½-inch howitzers; and

* Nouveau Système d'Artillerie de Campagne de Louis Napoleon Bonaparte, &c., par le Capitaine Favé, p. 11.

† Ibid., p. 12.

besides these varieties of calibre, and, as a matter of course, ammunition also for the artillery, six kinds of small-arm cartridges were carried into the field.*

Early use of the howitzer in India.

The howitzer gun which now forms such an important part of the British as well as the continental field equipments, appears in its earliest state to go back to A. D. 1480, when shells were fired into the palace of the rajah of Champanier,† possibly by such pieces as those seen by the writer at Boorhampoor.‡

The colossal Indian gun, which has long attracted attention in front of the officers' barracks at Woolwich, seems to belong to the class of howitzers, and does not differ essentially from those used in Germany in the sixteenth century. The howitzers, called feuerkatzen, were about 6 feet long, and had bores varying between 6 inches and 10 inches diameter. Pieces of this kind are mentioned A. D. 1544. Others of a larger size were used by the Emperor Ferdinand against the Turks in 1556, and by the King of Poland in the campaign of 1557.§ A little later (1605) hollow shot appear in "L'Instruction sur le fait de l'artillerie dressée, par le Duc de Sully."‖

Adopted in Germany in the 16th century.

* Remarks on the Organization of the Corps of Artillery in the British Service, pp. 30, 31. Rowland Hunter, 1818.

† Ibid., p. 32. ‡ Ibid., p. 36.

§ Against the Eifflenders.—Senfftenberg, tome vii., p. 353.

‖ Etudes sur le passé et l'avenir de l'Artillerie, par le Prince Louis Napoleon Bonaparte, tome i., pp. 166, 167, 212, 266, 267.

European powers appear to have made the mistake of adopting the short in preference to the long howitzers, with the exception of the Turks, from whom General Osmolski, of the Polish army, captured one of the longer and more efficient weapons in 1745. It is understood that Tomanowicz, a Hungarian general, used one of these pieces about 1765; and during the struggle previous to the partition of Poland in 1772 the Russians took some howitzer guns, which they adopted under the name of unicorns. A Turkish piece in the Repository at Woolwich, nearly corresponding with our 24-pounder howitzers, bears the date of 1805, when many others were cast by order of Sultan Selim.

Used by the Russians as unicorns.

A battery of unicorns was taken from the Russians by the French after it had done good service at the battle of Smolensko. Napoleon, on seeing the captured guns next morning, is said to have exclaimed, "Ce sont ces diables-là qui nous ont attrapés de si loin hier." An improved instrument of this kind was the consequence of their attracting Napoleon's attention; and the celebrated Paixhans' gun, which now takes a prominent place both by land and sea, subsequently appeared. On one of these, and that almost the smallest calibre, has been based the new system of field artillery by Louis Napoleon Bonaparte, according to which a 12-pounder howitzer gun is

Prince Louis Napoleon's system of artillery.

hereafter to serve for every purpose in the field. The ideas of the prince have been published, in order that artillery officers* and other military men may have an opportunity of deciding upon its merits; and it is understood that for this purpose the results have been given of at least some of those questions which were to have been discussed in the unpublished volumes "du passé et de l'avenir de l'artillerie."

12-pounder howitzers to serve for all purposes.

Taking "augmentation de simplicité, de mobilité, d'efficacité," as the principles to be carried out, he endeavours, by the use of a single piece, to remedy the inconveniences which may arise from the use of four calibres, namely, 12-pounder and 8-pounder guns, and 16-pounder and 15-pounder howitzers. When, for instance, a rapid movement of the reserve batteries becomes necessary to carry a particular point, the heavy guns only may chance to be available at the moment. Again, a rapid advance to complete the route of a retreating enemy may be foiled by his occupying a post at the critical moment when only 8-pounders are at hand to reduce it. But inconveniences such as these during successful operations are far less serious than those which may be experienced during a reverse, when some batteries, which may have exhausted the ordinary

* Nouveau Système d'Artillerie de Campagne de Louis Napoleon Bonaparte, &c., par le Capitaine Favé, &c. Paris, J. Dumaine, 1851.

supply, find, owing to certain caissons having been abandoned, that the guns and howitzers are without suitable ammunition; and this at the very instant when, although, perhaps, other ammunition is at hand, it is useless from the want of the pieces for which it was intended. Owing to such a mischance the guns may be without the means of doing good service, and thus become inert, and useless incumbrances. Feeling the necessity of providing for such contingencies, and bearing in mind the observation of the great warrior his uncle, that "Gribeauval a beaucoup reformé, il a beaucoup simplifié; mais l'artillerie est encore trop lourde, trop compliquée; il faut encore simplifier, uniformer, réduire jusqu'a ce qu'on soit arrivé au plus simple,"* the Prince proposes to confine the field service to 12-pounder howitzer guns, mounted on an 8-pounder carriage, and retaining only four projectiles, viz., shells, solid shot, ordinary case and spherical case shot, using for these smaller charges than heretofore. A diminished charge is based on the fact, that at the distance of 1,000 yards the penetration of a 12-pounder with a charge of ¼ of its weight is only 3 centimetres or 1·38 inches less than that of a 12-pounder discharged with ⅓ of its weight; and a larger calibre was proposed in consequence of finding that a 12-pounder shot

Advantages of this change.

Simplification of artillery recommended by the Emperor Napoleon.

* Mémoires de Napoleon, &c., tome i., p. 271.

with ¼ of its weight of powder penetrated farther at every distance than an 8-pound shot with a charge of ⅛ of its weight, and that it was at the same time more accurate.*

Relative weights of 12-pounder or 8-pounder guns.

By using a charge of ¼ the weight of the shot, that of the gun can be reduced to 659 kilogrammes, or 1,344 lbs. (French); that is, about 161 French lbs. more than the French 8-pounder gun; and, being mounted as proposed on the carriage of the latter, the weight of both is 2,427 lbs.†

TABLE No. 1.—PRESENT AND PROPOSED SYSTEM.

TABLE showing the WEIGHT of the GUNS, that of their CARRIAGES, and also of the GUNS with their CARRIAGES, and the Number of HORSES to each.

Weight of the old and new guns.

—	Weight of the Pieces.	Weight of the Carriages.	Weight of the Gun with its Carriage and Limber.	Number of Horses.
	lbs.	lbs.	lbs.	
12-pounder gun . . .	1800	1165	4410	6
8 ,, ,, ...	1183	1083	3694	6
16-pounder howitzer .	1810	1165	4433	6
15 ,, ,, .	1185	1083	3743	6
12-pounder howitzer gun	1344	1083	3800	6

Artillery and the new

NOTE.—The waggons when loaded weigh between 200 lbs. and 700 lbs. less than the pieces to which they belong. They are drawn by six horses in the batteries, and four horses in the parks. The 24-pounder on its travelling carriage has eight horses, and so has the 16-pounder.‡ The French weights are given, ammunition included.

With respect to that point on which it is admitted that there is some inferiority, viz., shell-

* Nouveau Système d'Artillerie de Campagne de Louis Napoleon Bonaparte, &c., par le Capitaine Favé, pp. 14, 15, 17.

† Ibid., pp. 21–45. ‡ Ibid., p. 45.

firing, it is presumed that a certain weight of projectiles thrown by a 12-pounder howitzer gun would produce quite as great an effect as a certain number of 15 and 16 pounder shells of equal weight. The efficiency of 12-pounder shells was established by experiments made in Austria, when 100 shells destroyed the epaulement of a battery so completely, that its three guns were entirely exposed in consequence; nor are the effects of its projectiles, when used as carcasses, inferior (in proportion) to those of the larger shells.

12-pounder shells compared with 15 and 16 pounder shells.

"But the main object," says Captain Favé, "is so to increase the efficiency of modern artillery, that it may keep pace with improved musketry." Since the last war, percussion fire-arms have been introduced; and recently, as has been seen,* a particular kind of carbine, with greatly increased power as to precision and range. Every possible effort should, therefore, be made to improve the artillery in the same proportion. It is to be hoped, though experience alone can decide the point, that the proposed 12-pounder gun will be superior to the howitzers hitherto in use, and also have the advantage of dispensing with three of the projectiles now used in field equipments. A shell filled with balls weighs as much as a solid shot, and acquires as much velocity; consequently, the

Artillery should keep pace with the improved musket.

* Nouveau Système d'Artillerie de Campagne de Louis Napoleon Bonaparte, &c., par le Capitaine Favé, p. 30.

The 12-pounder howitzer gun compared with 12 and 8 pounder guns.

new instrument will accomplish the same object as an ordinary cannon of the same calibre, thus performing effectually the work of two different pieces of ordnance.* And whilst the proposed howitzer will have rather more rounds of ammunition than the 12-pounder and almost as many as the 8-pounder gun, both of which it is intended to replace, it will have nearly the same lightness and facility of movement as belong to the latter.

The 12-pounder howitzer gun will replace four different calibres.

Should the new system of artillery be adopted, the four calibres and nine kinds of projectiles now in use will be replaced by one kind of gun, which, on the whole, has an advantage over the four pieces in question in point of range, accuracy, and penetration, with, at the same time, the greatly-to-be-appreciated advantage of discharging at pleasure from every piece, either solid shot, spherical case, or ordinary shells. Even two of the new pieces may compete with a 12-pounder and an 8-pounder; and the comparison is still more favourable when there is a greater number. According to the existing proportion, there is but one 12-pounder to five 8-pounder batteries. Therefore if there should be six batteries of each kind, the howitzer guns would discharge thirty-six 12-pounder shot, whilst those armed according to the existing system would fire only four

Missiles discharged by the new piece.

* Nouveau Système d'Artillerie de Campagne de Louis Napoleo Bonaparte, &c., par le Capitaine Favé, pp. 33, 34.

12-pounders, twenty 8-pounders, two 16-pounders, and ten 15-pounder shells. and their comparative effects.

The new idea of the President of the French republic, says Captain Favé, must bring some inconveniences, as was the case even with the valuable improvements of Gribeauval on their first introduction but the essential point is to ascertain whether, *on the whole*, the advantages preponderate. In order to decide this question, detailed experiments were ordered, and carried on simultaneously at Metz, Strasbourg, Toulouse, and Vincennes, to ascertain the relative advantages of the new howitzer gun and the present field-pieces, viz., 12 and 8 pounder guns, and 16 and 15 pounder howitzers. The new piece tested at Metz, Strasbourg, &c.

The commissions were to examine the following points:—

Objects of the experiments.

1. Accuracy of fire of shot and shell.
2. Their penetration into the earth.
3. The effects of the fire of shot.
4. Recoil of the carriages.
5. Durability of the carriages and the effect of the firing, comparing, in the first instance, the howitzer with the 12-pounder and 8-pounder guns, and again with the 16-pounder and 15-pounder howitzers, under the various circumstances which are detailed. The commission was also instructed to make a report of the result of the experiments as regards a battery of six

Batteries of 12 and 8 pounders compared.

howitzer guns compared with one of the ordinary 12-pounders and another of 8-pounders. The results of experiments at the various distances of 640·5, 768·6, 896·7, 1,024·8, and 1,152·9 yards, which took place independently at the several places in question, are given in the following table, and also the average per centage of the whole :*—

TABLE No. 2.—PRACTICE WITH SOLID SHOT.

NUMBER of SHOTS FIRED, and also the Number which took effect, on a Screen or Target 32·8080 yards wide by 3·2808 yards high.

Commissions.	Pieces.	Charge.	Shots.	Distances—Yards. 546·8165	656·1798	765·5431	874·9064	984·2697	Total of Shots at the Five Distances.
		Kilogram.		No. of rounds fired.					
TOULOUSE.	12-pounder gun	1,958	Fired . .	42	42	42	42	42	210
			Taking effect	28	21	19	14	16	98
	8 ,,	1,223	Fired . .	39	39	39	39	40	196
			Taking effect	27	20	15	10	15	87
	12-pr. howitzer gun . . .	1,500	Fired . .	36	36	36	37	37	182
			Taking effect	20	23	16	11	16	86
VINCENNES.	12-pounder gun	1,958	Fired . .	42	42	42	42	42	210
			Taking effect	25	20	15	14	12	86
	8 ,,	1,223	Fired . .	39	39	39	39	40	196
			Taking effect	29	15	19	9	10	82
	12-pr. howitzer gun . . .	1,500	Fired . .	36	36	36	37	37	182
			Taking effect	28	17	14	16	11	86
STRASBOURG.	12-pounder gun	1,958	Fired . .	42	42	42	42	42	210
			Taking effect	25	26	21	12	10	94
	8 ,,	1,223	Fired . .	39	39	39	39	40	196
			Taking effect	23	15	13	16	12	79
	12-pr. howitzer gun . . .	1,500	Fired . .	36	36	36	37	37	182
			Taking effect	27	19	16	17	10	89

* Nouveau Système d'Artillerie de Campagne de Louis Napoleon Bonaparte, &c., par le Capitaine Favé, pp. 61, 62.

TABLE No. 2.—PRACTICE WITH SOLID SHOT—*continued.*

Commissions.	Pieces.	Charge.	Shots.	Distances—Yards. 546·8265	656·1798	765·5431	874·9064	984·2697	Total of Shots at the Five Distances.
		Kilogram.		No. of rounds fired.					
METZ.	12-pounder gun	1,958	Fired	42	42	42	42	42	210
			Taking effect	30	25	18	23	16	112
	8 ,,	1,223	Fired	39	39	39	39	40	196
			Taking effect	26	19	16	10	9	80
	12-pr. howitzer gun	1,500	Fired	36	36	36	37	37	182
			Taking effect	25	19	21	14	12	91
	Result of the Four Commissions united.								
	12-pounder gun	1,958	Fired	168	168	168	168	168	840
			Taking effect	108	92	73	63	54	390
	8 ,,	1,223	Fired	156	156	156	156	160	784
			Taking effect	105	69	63	45	46	328
	12-pr. howitzer gun	1,500	Fired	144	144	144	148	148	728
			Taking effect	100	78	67	58	49	352
	Mean Result of the Four Commissions, showing the numbers of Shots per Cent.								
	12-pounder gun	1,958	. .	64·2	54·4	43·4	37·5	32·1	46·4
	8 ,,	1,223	. .	67·3	44·2	40·3	28·8	28·7	40·5
	12-pr. howitzer gun	1,500	.	69·4	54·1	46·5	39·1	33·1	48·3

N.B. The ranges and size of the targets are given in yards to facilitate the examination.

Advantages of the 12-pounder howitzer gun.

It will thus be seen that the howitzer gun, when firing a solid shot with a charge of one-fourth of its weight, had the advantage over the 12-pounder gun in the proportion of 48 to 46 and of 33 to 32 per cent. at every one of the ranges, except at 768·6 yards, when it was equal. But compared with the 8-pounder, the result was still more favourable, being about 33 to 28.

The following table gives the lateral deviations :—

Lateral deviations of the shot.

TABLE No. 3.

PRACTICE with SOLID SHOT, and Mean Lateral Deviation, in Yards.

Commissions.	Pieces.	Charge.	Distances—Yards.					Mean of the Deviations at the Five Distances.
			546·8165	656·1798	765·5431	874·9064	984·2697	
		Kilogrammes.	Yards.	Yards.	Yards.	Yards.	Yards.	Yards.
VINCENNES	12-pounder gun . .	1,958	1·7278	2·1434	2·8324	3·8713	3·0839	2·7230
	8 ,, ,, . .	1,223	1·1920	2·1653	2·7340	4·4181	4·5384	3·0074
	12-pounder howitzer gun	1,500	1·7825	2·3731	2·2528	3·8604	4·1556	2·8761
STRASBOURG	12-pounder gun . .	1,958	0·9295	1·8919	1·8372	2·4387	2·8761	1·9794
	8 ,, ,, . .	1,223	1·4872	1·4544	2·9855	2·8324	2·9855	2·3621
	12-pounder howitzer gun	1,500	1·2357	1·6950	2·2309	2·5262	3·5760	2·5152
TOULOUSE	12-pounder gun . .	1,958	1·3560	1·7935	1·8263	3·4557	3·6197	2·3277
	8 ,, ,, . .	1,223	1·8153	2·1762	2·6574	3·8604	4·4618	2·9964
	12-pounder howitzer gun	1,500	1·8312	2·3184	3·1386	2·6355	3·6197	2·7121
METZ	12-pounder gun . .	1,958	1·5877	1·8263	1·6513	2·4824	4·0791	2·3293
	8 ,, ,, . .	1,223	1·8312	2·1215	2·0012	4·1119	4·5493	2·9119
	12-pounder howitzer gun	1,500	1·7169	1·4654	2·5590	3·2261	4·1010	2·5037
	Mean Result of the Four Commissions.							
	12-pounder gun . .	1,958	1·3998	1·9138	2·0340	3·0411	3·4120	2·3621
	8 ,, ,, . .	1,223	1·5857	1·9794	2·5917	2·7121	4·1338	2·8214
	12-pounder howitzer gun	1,500	1·6404	1·9575	2·5590	3·0620	3·8604	2·7230

The mean results of the four different experiments show that the shots discharged by a 12-pounder howitzer take intermediate places between those of the 12-pounder and 8-pounder gun; the latter, however, having a slight advantage in this respect as regards shot. The following tables give the result of the experiments made with shells, including their lateral deviations:—

Experiments with shells.

TABLE No. 4.—HOWITZER PRACTICE.

NUMBER of SHOTS FIRED, and also the Number which took effect, on a Screen or Target 32·8080 yards wide by 3·2808 yards high.

Commission.	Pieces.	Charge.	Shots.	Distances—Yards. 546·8165	656·1798	765·5431	874·9064	984·2697	Total.
				No. of rounds fired.					
TOULOUSE.	16-pounder howitzer	Full charge .	Fired . .	. .	. .	. .	13	13	26
			Taking effect	. .	. .	. .	5	8	13
		Reduced charge	Fired . .	26	26	26	14	14	106
			Taking effect	9	7	7	3	3	29
	15 ,,	Full charge .	Fired . .	. .	. .	. .	13	13	26
			Taking effect	. .	. .	. .	. .	2	2
		Reduced charge	Fired . .	28	28	28	15	15	114
			Taking effect	9	8	3	2	. .	22
	12-pr,hr.gun	Charge 1 K 225	Fired . .	36	36	36	37	37	182
			Taking effect	24	30	21	11	15	101
VINCENNES.	16-pounder howitzer	Full charge .	Fired . .	. .	. .	. .	13	13	26
			Taking effect	. .	. .	. .	3	1	4
		Reduced charge	Fired . .	26	26	26	14	14	106
			Taking effect	14	5	5	1	3	28
	15 ,,	Full charge .	Fired . .	. .	. .	. .	13	13	26
			Taking effect	. .	. .	. .	5	1	6
		Reduced charge	Fired . .	28	28	28	15	15	114
			Taking effect	10	4	4	1	. .	19
	12-pr.hr.gun	Charge 1 K 223	Fired . .	36	36	36	37	37	182
			Taking effect	20	19	11	10	8	68

TABLE No. 4.—HOWITZER PRACTICE—*continued.*

Commissions.	Pieces.	Charge.	Shots.	Distances—Yards. 546·8165	656·1798	765·5431	874·9064	984·2697	Total.
				No. of rounds fired.					
STRASBOURG.	16-pounder howitzer	Full charge	Fired . .	..	..	..	13	13	26
			Taking effect	..	..	..	5	6	11
		Reduced charge	Fired . .	26	26	26	14	14	106
			Taking effect	13	9	8	2	..	32
	15 ,,	Full charge	Fired . .	..	..	..	13	13	26
			Taking effect	..	..	..	5	2	7
		Reduced charge	Fired . . .	28	28	28	15	15	114
			Taking effect	13	8	4	3	..	28
	12-pr.hr.gun	Charge 1 K 225	Fired . . .	36	36	36	37	37	182
			Taking effect	23	26	10	13	10	82
METZ.	16-pounder howitzer	Full charge	Fired . . .	..	..	..	13	13	26
			Taking effect	..	..	..	7	4	11
		Reduced charge	Fired . . .	26	26	26	14	14	106
			Taking effect	14	7	5	2	2	30
	15 ,,	Full charge	Fired . . .	..	..	..	13	13	26
			Taking effect	..	..	..	2	4	6
		Reduced charge	Fired . . .	28	28	28	15	15	114
			Taking effect	9	5	8	2	..	24
	12-pr.hr.gun	Charge 1 K 225	Fired . . .	36	36	36	37	37	182
			Taking effect	30	27	20	18	16	111
	Results of the Four Commissions united.								
	16-pounder howitzer	3·3081	Fired . .	..	..	..	52	52	104
			Taking effect	..	..	..	20	19	39
		1·1541	Fired . . .	104	104	104	56	56	424
			Taking effect	50	28	25	8	8	119
	15 ,,	2·2054	Fired . . .	..	..	..	52	52	104
			Taking effect	..	..	..	12	9	21
		1·1027	Fired . . .	112	112	112	60	60	456
			Taking effect	11	25	19	8	..	93
	12-pr.hr.gun	2,7016	Fired . .	144	144	144	148	148	728
			Taking effect	97	102	62	52	49	362
	Mean Result per Cent. of the Four Commissions.								
	16-pounder howitzer	1 K 50	. .	..	..	..	38·4	36·4	37·5
		0 K 75	. .	48·0	26·9	24·0	14·2	14·2	28·0
	15 ,,	1 K 0	. .	..	..	..	23·0	17·3	20·3
		0 K 50	. .	36·6	22·3	16·9	13·3	..	20·4
	12-pr.hr.gun	1 K 225	. .	67·3	70·8	43·0	35·0	33·0	49·7

TABLE No. 5.—HOWITZER PRACTICE. MEAN LATERAL DEVIATION of the SHELLS, in Yards.

Commissions.	Pieces.	Charge.	Distances—Yards.					Mean of the Deviations.
			546·8165	656·1798	765·5431	874·9064	984·2697	
			Yards.	Yards.	Yards.	Yards.	Yards.	Yards.
VINCENNES	16-pounder howitzer	Full charge .	. .	. .	. .	2·8105	1·8919	2·3512
		Small charge	3·0074	4·2650	4·2322	6·1122	11·8874	5·9008
	15 ,, ,,	Full charge .	. .	. .	. .	3·9041	4·8227	4·3634
		Small charge	4·3197	4·0135	4·7680	10·5969	32·2611	11·1016
	12-pr. howitzer gun.	. .	1·3888	1·9356	3·5870	3·6307	4·3415	2·9767
STRASBOURG	16-pounder howitzer	Full charge .	. .	. .	. .	2·1543	5·5008	3·7340
		Small charge	2·6246	4·0463	5·9054	6·0813	7·6661	5·2647
	15 ,, ,,	Full charge .	. .	. .	. .	3·1605	5·5117	4·3340
		Small charge	3·1167	4·0353	8·1363	4·9102	13·0466	6·6490
	12-pr. howitzer gun.	. .	1·4216	2·7449	3·3464	3·6307	6·4631	3·5213
TOULOUSE	16-pounder howitzer	Full charge .	. .	. .	. .	2·8214	6·0366	4·4145
		Small charge	2·4606	3·6307	5·0414	9·7002	8·0598	5·7785
	15 ,, ,,	Full charge .	. .	. .	. .	6·9054	5·5554	5·7304
		Small charge	2·8652	5·2820	6·8677	9·6346	7·1740	6·3647
	12-pr. howitzer gun.	. .	1·6185	2·1553	3·5323	4·2541	6·1241	3·5388
METZ . .	16-pounder howitzer	Full charge .	. .	. .	. .	2·4277	3·9807	3·2042
		Small charge	2·6246	3·2479	3·5323	4·9649	6·3100	4·1359
	15 ,, ,,	Full charge .	. .	. .	. .	3·3016	4·3196	3·8106
		Small charge	3·1386	4·6149	5·1508	6 4194	10·5531	5·9754
	12-pr. howitzer gun.	. .	1·3013	2·8542	2·0549	2·4168	4·5821	2·6404
Mean Result of the Four Commissions united	12-pounder howitzer	Full charge .	. .	. .	. .	2·5480	4·3525	3·9502
		Small charge	2·6793	3·7947	4·6806	6·7147	8·4754	5·2689
	15 ,, ,,	Full charge .	. .	. .	. .	4·0681	5·0524	4·5603
		Small charge	3·3573	4·4837	6·2225	7·8846	15·7587	7·5413
	12-pr. howitzer gun.	. .	1·4326	2·4277	3·1276	3·8057	5·3805	3·1748

Comparison of the 12-pounder with 15 and 16 pounder howitzer.

Table 4 will show that, taking a mean of all the distances, the howitzer gun has the advantage not only over the 15-pounder howitzer, but over the 16-pounder also. The lateral deviation of the 12-pounder howitzer, as shown by Table 5, is greater than that of the 16-pounder howitzer fired with a full charge. It is equal, or nearly so, to the 15-pounder using its full charge; and is superior to both the 16 and 15-pounder howitzer, when using small charges.

The more important consideration of power, which in the case of the howitzer might be supposed to be greatly inferior to the ordinary cannon, is shown with regard to solid shot by the following table:—

Mean penetration with solid shot of the old and new gun.

TABLE No. 6.

MEAN PENETRATION of SOLID SHOT into RAMMED or SOLID EARTH, in Yards.

Commissions.	Pieces.	Charge.	Distances.	Mean Penetration
		lbs.	Yards.	Yards
VINCENNES	12-pounder gun	4·3181	32·8080	1·4107
	8 ,, ,,	2·6972	32·8080	1·2029
	12-pr. howitzer gun	3·3081	32·8080	1·2685
STRASBOURG	12-pounder gun	4·3181	21·8720	1·9903
	8 ,, ,,	2·6972	21·8720	1·7825
	12-pr. howitzer gun	3·3081	21·8720	1·9138
TOULOUSE	12-pounder gun	4·3181	32·8080	1·4107
	8 ,, ,,	2·6972	32·8080	1·1920
	12-pr. howitzer gun	3·3081	32·8080	1·1592
METZ	12-pounder gun	4·3181	38·2760	2·1434
	8 ,, ,,	2·6972	38·2760	1·7935
	12-pr. howitzer gun	3·3081	38·2760	1·9138

And in the case of howitzers by another:—

TABLE No. 7.

MEAN PENETRATION of SHELLS into RAMMED or SOLID EARTH.

Commissions.	Pieces.	Distances.	Charge.	Depth.	Charge.	Mean Depth.
		Yards.	lbs.	Yards.	Kilogram.	Yards.
VINCENNES	16-pounder howitzer	87·4880	3·3081	1·2248	0·75	1·1045
	15 ,, ,,	87·4880	2·2054	1·0389	0·50	0·9842
	12-pr. howitzer gun	87·4880	2·7016	0·9295	1·225	0·9076
STRASBOURG	16-pounder howitzer	109·3600	3·3081	1·6950	0·75	1·3341
	15 ,, ,,	109·3600	2·2054	1·3123	0·50	1·1373
	12-pr. howitzer gun	109·3600	2·7016	1·2138	1·225	1·2138
TOULOUSE .	16-pounder howitzer	54·6800	3·3081	1·1264	These were not fired with small charges.	
	15 ,, ,,	54·6800	2·2054	0·9951		
	12-pr. howitzer gun	218·7200	2·7016	6·999		
METZ . .	16-pounder howitzer	109·3600	3·3081	1·8809	0·75	1·7497
	15 ,, ,,	109·3600	2·2054	1·5638	0·50	1·2357
	12-pr. howitzer gun	109·3600	2·7016	1·6732	1·225	1·6075

None of the 12-pounder shells burst in the pieces or during the ricochet along the ground; but, as might have been expected at such short distances, some of the shells were broken by the concussion of the earth. This occurred at Vincennes, also at Strasbourg and Toulouse, but not at Metz, although firing at the short distance of 100 metres. It will be seen by the preceding tables that the mean penetration of the howitzer gun, when using solid shot, was greater than that of the ordinary 12-pounder as well as of the 8-pounder, both at Vincennes and Strasbourg; whilst at Toulouse the 12-pounder howitzer was rather inferior to guns of the same calibre, but was as nearly as possible equal to the 8-pounder,

Mean penetration of solid shot into solid earth.

Experiments favourable as to the carriage of the new gun.

which it surpassed at Metz. There was, as might have been expected, a considerable variety in the recoil of the different pieces during the experiments; but that of the howitzer gun, whether using solid shot, case shot, or shells, was within the limits of the other howitzers and guns. Moreover it was less severe on its carriage than the other pieces.

Efficiency of batteries of the new gun,

In order to show the greater efficiency of the howitzer gun as the result of the experiments given in Tables 2, 3, 4, and 5, let us compare six batteries as at present organized, with a like number of howitzer guns:—

Of the	1,137	shots discharged by a 12-pounder reserve battery	481	take effect.
And of the	5,560	shots discharged by 8-pounder batteries of division . . .	1,940	,,
Total	6,697		2,421	taking effect.

Of the	1,488	shots discharged by a 12-pounder reserve battery of howitzer guns	728	take effect.
And of the	5,280	shots discharged by a 12-pounder battery of division	2,585	,,
Total	6,768		3,313	taking effect.

Giving, as a mean result of experiments at five different ranges, a difference of 892, or more than one-third in favour of the new arrangement. And if the whole discharges had taken place at the longest of the preceding ranges, namely,

900 yards, the result would have been still more decided :—

Of a 12-pounder reserve battery . . .	326	shots take effect,
and of five 8-pounder batteries of division	1,235	,,
or . . .	1,561	shots for the six of the existing batteries,
and . . .	2,231	shots for the proposed batteries,

being a difference of 670 in favour of the latter. With respect to the effect against substantial objects, such as a wall of 3,843 yards long by 3,843 yards high, the aggregate force of six of the existing batteries would be represented by 891, whilst the same number, armed as proposed, would give 1,058, or a difference of 167 in favour of the latter; and 120 shells will have taken effect of the former to 1,115 of the latter, or nine times the number;* the splinters or fragments being, as a matter of course, proportionably greater.

compared with those of the former pieces.

The principal objections which have been started by the several commissions are: 1st. That two kinds of field batteries are desirable, and even necessary; and that a great moral effect is produced when the reserve batteries are brought up. It is, however, observed in reply, that in the first instance an enemy cannot perceive the difference between 12-pounders and 8-pounders; and if the new battery be nearly equal in power to the

Objections offered to the new system,

* Nouveau Système d'Artillerie de Campagne de Louis Napoleon Bonaparte, &c., par le Capitaine Favé, pp. 80–87.

former, it is manifestly superior, and must be preferable to the 8-pounders.

2ndly. That a battery of howitzer guns, in addition to being equally moveable, should have as much ammunition as a battery of 8-pounders, in which latter particular, it fails.

and the advantages set forth in reply.

It is admitted, in reply, that the proposed battery of division would have rather less ammunition than 8-pounders; but as a greater number of shots take effect in proportion, there would be a greater number of effective discharges in the same time. Moreover, when a comparison is made with six batteries of various calibres, there would be an equal number of rounds and more weight propelled by the howitzers.

3rd objection. Shot thrown by the howitzer gun at a long range has neither the power nor the accuracy of fire of the 12-pounder gun.

With solid shot the howitzer has equal accuracy.

It is true that both the 12-pounder and 8-pounder have a more direct fire than the howitzer gun; but at 1,153 yards the latter has the advantage in point of precision, although having the disadvantage of being thrown at a higher angle; and as its velocity is better preserved in proportion, it has the advantage in more distant ranges. In fact, the fire of the howitzer gun with solid shot is on the whole more accurate than that of the guns in question, viz., the 12 and 8 pounders.

4th objection. A small hollow shot cannot advantageously replace the shells in use, the power of which depends on their calibre; and the 12-pounder can have little effect at long distances. Supposed disadvantage of smaller shot,

With reference to the first part of the objection, it must be admitted that the 12-pounder howitzer gun has not as much power to overcome stone walls as the large shells; but to its greater precision it will, weight for weight, have greater effect than a certain number of ordinary shells, which, when fired with full charges, would either be broken against, or miss the wall.

With regard to the latter part of the objection, when long ranges are in question, solid shot from the proposed gun must be superior to an 8-pounder, without being much inferior to the ordinary 12-pounder.

5th objection. According to the new project, the number of shells would be too great.

If in one sense this objection be well founded, it is because the fire has become greatly more effective. Between 640·5 yards and 1,152·9 yards, the proposed cannon is more accurate than the 8-pounder gun, and being of the same weight it may be substituted for it, with the advantage of doing execution as a shot, as well as by its splinters when it bursts. and bearings of this question.

6th objection. The present reserve batteries

having more shot, are capable of producing a more decided effect.

General result of the comparisons.

It must be admitted that in the case of the ordinary 12-pounder batteries, 399 shells take effect out of 861; and but 359 out of 744 in a howitzer battery, or 40 less. Therefore, under the rare circumstances of firing against walls, the proposed reserve battery would be inferior to the 12-pounder. This, however, is more than compensated by the advantages to be derived from the division howitzer-gun batteries, and the other reserve batteries of this kind, which, it will be remembered, discharge 12-pounder instead of 8-pounder shot.

Relative power of carcasses.

7th objection. When used as a carcass, the 12-pounder is inferior to the 15 and 16 pounder howitzer shells.

It should be observed, in reply, that the existing batteries require the use of the 15-pounder as well as the 16-pounder shells to set fire to buildings; whereas the 12-pounder, owing to the number discharged, takes the place of both, with the additional advantage of having greater precision.

Expense and other objections to

8th objection. Finally, the proposed change would involve considerable expense by doing away with the existing material.

To this it may be answered, that of the 5,600 pieces of ordnance now in use, 3,800 guns and howitzers will be required, with their shot and shell, for the different fortresses; so that only about 1,800 howitzer guns will have to be recast to complete the field equipment with one kind of gun: and this might be accomplished in the course of about twelve months, at an expense, including the transport to different places, of a sum not exceeding 2,000,000 francs.*

the new system more than counter-balanced by

Having met the preceding as well as other objections of less moment, Captain Favé thus concludes his exposition of the new system of field artillery by Louis Napoleon Bonaparte:—

"We repeat that there can be no improvements so perfect as not to give cause for criticism. But, with the exception of some trifling and even doubtful inconveniences, we perceive that the proposed system has the clearest and most appreciable advantages during a campaign in the field, owing to the manifest simplification of the equipments, the increased precision and power of fire, and the greater durability of the carriages." †

its simplicity and other considerations.

The use of a howitzer gun in our service has not been by any means neglected, one-half of the horse artillery armament being of this description; and in case of the necessity which might arise of reducing a mill or other substantial building, a 24-pounder howitzer is attached to each of the field batteries in Ireland. Doubtless,

The howitzer gun, which is partially used,

* Nouveau Système d'Artillerie de Campagne de Louis Napoléon Bonaparte, &c., par le Capitaine Favé. Substance of the objections and answers, as given pp. 88–127.

† Ibid., p. 128.

therefore, the larger question which now occupies the French artillery officers, will receive, if it has not already done so, the same painstaking consideration which has been given to the Delvigne carbine and the other new muskets. Such a change in artillery as that proposed is a large and momentous question, the result of which may possibly be the future use of only two kinds of field-pieces in the British service, viz., for horse and light artillery a howitzer gun of about the weight of a light 6-pounder, and for the batteries a 12-pounder howitzer gun of rather less weight than the present 9-pounder, and adapted equally for solid shot and shells.

to become general in the British army.

A change in the arms of infantry,

The writer has not ventured to express all that he feels regarding the proposed simplification of field artillery, and the superiority that might be maintained in consequence over the new small arms; but as the latter are about to be introduced in our army, there is less reason for objecting to a corresponding change in another branch of the service. The march of improvement in science is retarded by two serious enemies—imprudent changes, and a still more formidable adversary, routine.* If, after due consideration, and a full examination, something of this kind should promise corresponding advantages, the intended sub-

may be followed by one in the artillery.

* Avant propos, tome i., p. xi., d'Etudes sur le passé et l'aveuir de l'Artillerie, par le Prince Louis Napoleon Bonaparte.

stitution of 9 for 6 pounders, offers a favourable opportunity for the introduction of the 12-pounder howitzer gun, since, having rather less metal, one description of 12-pounder howitzer gun might, with very little additional expense, replace the present 9-pounder, whilst another, about the weight of the 6-pounder, would be substituted for the latter description of gun.

Deficiency and other disadvantages of this service.

If the pains bestowed by the writer on this subject have not been altogether lost, the preceding pages will have shown that some disadvantages still belong to the British artillery; amongst which, in addition to the admitted and inherent vices of want of promotion and age of the senior officers, the following are paramount:—

That the artillery, whether serving in the colonies, at home, or in the field, does not bear a due proportion to the other two arms. That in the latter case, as compared with the number of guns taken into the field in 1742 and 1762, we have been retrograding rather than the reverse;* or, to use the words of the hero of the Peninsular war, the proportion of artillery "was infinitely lower than that of any army acting in Europe of the strength of the British part of the allied army alone, and below the scale he had ever read of for any army of such numbers."†

* See above, pp. 122-126.

† Despatch to Earl Bathurst, dated Frenada, January 27, 1813.

Great deficiency of artillerymen

From the rough estimate made at pages 111, 112, and 113, it will be seen that, in addition to the present field equipments, something like 3,000 additional gunners would be absolutely necessary to serve a due proportion of artillery for the militia and volunteer force that would be placed under arms in case of a sudden emergency. Being without horses, this would be taking almost inert masses into the field, or at best mere guns of position, which would be almost useless, unless the majority of the men should be expert gunners. Supposing nearly half of the men attached to the guns to be furnished from the dock-yards, the line, and other sources, about 1,500 artillerymen would still be required in addition, without taking into account those troops who might chance to be on an expedition at the moment on some part of the Continent.

in case of an invasion.

More artillery employed in 1748 in Flanders

Looking back to the instances already mentioned,* of the number of guns with the army in Germany and in Flanders at various times, we find that there were 78 pieces of ordnance, with 22 battalions and 14 squadrons employed in the latter country in 1748; and as the force in question could scarcely have exceeded 20,000 men, we see that the artillery at that time greatly exceeded the proportion with the Duke of Wellington's army when England was engaged in

* Pp. 122–126.

the most vital struggle. At Salamanca, for instance, there were but 24 guns to 20,000 men; whilst, as observed by his Grace, "the French had more than twice that number in action."* — than in Spain in 1812.

The author has not, however, ventured to give more than an idea of the extent to which the increase of the artillery should be carried, having contented himself with endeavouring to show that the proportion of artillery is lamentably deficient, and that some such reorganization as that which has been proposed is absolutely necessary, in order that the horse, field, and garrison artillery may be considered as a whole, and be not only in proportion to the line, but also be adapted for the various exigencies of the British empire. — The artillery service should be considered as a whole.

The writer has endeavoured to show that, in order to secure efficiency, it is absolutely necessary that the companies should be permanently attached to the field batteries, to the same extent at least as is the case with the horse brigade. This would necessarily involve a fresh organization and separation into field and garrison artillery, with (if it shall be thought right to provide, in some measure, a due proportion for the wants of both services) a considerable increase to the regiment, and, above all things, the important improvement of doing duty by battalions. This would not be a change to anything new, but — Separation of the field service. — The regiment to do duty by battalions.

* Despatch from Villa Toro, dated October 18, 1812.

merely a recurrence to what was the system of our regiment in common with all other services, from 1763 until the confusion and difficulties occasioned by the commencement of the revolutionary war, and the rapid and unmethodical accession of stations, caused a departure from it. It continued in abeyance from this time till 1822, when, with a view to changing the tours by companies, the Duke of Wellington assembled some of those belonging to the same battalion at Gibraltar. But the number of companies, at that time ten in each battalion, showed that the latter were not suited to the foreign stations. Difficulties, therefore, were caused, which do not apply to the present organization, or to battalions such as those proposed by Sir Augustus Fraser in 1818, of which the colonels and field-officers were to have been an integral part. Therefore, if the present detached or company system is to be continued, it becomes a question worthy of consideration whether one of those grades may not be dispensed with.

Employment of the colonels-en-second,

But if, with reference to efficiency, battalion in preference to company duty should be adopted, the present colonels, or rather younger men in their places, will be absolutely necessary for the command of each, as part of a reconstruction that seems calculated to substitute bodily vigour for the worn-out frames of the senior officers of the corps.

When the anomalous separation shall cease by placing the whole force of Great Britain under the commander-in-chief, there is little fear that any army or portion of the army will be allowed to take the field without having sufficient artillery to support the operations that may be undertaken; and as a necessary consequence of ceasing to be separate branches, the officers of artillery and engineers will be considered eligible, and will make themselves fit for general employment, instead of being excluded from it as at present; by which means they will be enabled to make some return to the country for the education received, and the pay drawn since their first appointment to commissions.

and general employment of the then officers.

In concluding these observations, it may here be briefly mentioned that, for reasons already given, the writer is strongly of opinion that horse artillery should, as at present, form a part of any future organization of the British artillery.

The horse artillery to be retained.

The guns of the latter and those of the field batteries being similar, it follows that their speed, when equally horsed, is, or may be, the same; but, as already observed, there is a substantial difference in the rest of the equipment, the gun detachments being mounted in the one case, and carried upon the waggons in the other, which are in consequence rather heavier than the guns to which they belong. This circumstance naturally

Comparison of the horse artillery

diminishes the speed of the field battery as compared with the horse artillery; but as it has been seen, p. 189, the difference is not by any means so great as to prevent the waggon from being made as light as the gun to which it belongs. Three gunners, mounted on the off-horses, as in the Bengal horse artillery, would of itself make the difference; and it might also be accomplished either by diminishing the weight of the carriage, or by lessening the quantity of ammunition. When of the same weight as the gun, the speed of the latter would be attained, and consequently that of the horse artillery also on level ground; and since the officers and men for each are furnished by the same corps, equal efficiency must be the

with the field batteries.

result of a permanent field service. But whilst the latter would be much less expensive, and have, under ordinary circumstances, other advantages, it must not be forgotten that, in broken ground and protracted exertion in difficult countries, the extra horses which could then be applied in draught, give to the horse artillery a decided advantage; and it is for this reason that some light guns, thus equipped, are indispensable.

But in venturing to make the preceding observations on an important and difficult subject, the author is far from desiring to see less care bestowed on the *élite* of the artillery; for he desires rather to say, in the words of our immortal bard, "Not Cæsar less, but Rome more."

Let the country, therefore, continue to cherish the horse artillery as before, but at the same time bestow the necessary attention to secure an efficient field artillery, and thus bring about that state of things so happily expressed by Captain Favé regarding the improvement made in both branches of the French service. His words are: —" Dans une place, le métier d'artilleur n'est pas soumis à des règles aussi fixes que dans la guerre de campagne; il exige plus de connaissances et d'intelligence. Nous avons fait précédemment remarquer que la grande extension donnée à l'artillerie à cheval avait nui à l'artillerie à pied, et que l'on eut souvent l'occasion de s'en apercevoir. L'histoire de l'artillerie doit relater et étudier avec soin de pareils faits, et cette arme doit s'efforcer d'éviter à l'avenir les mêmes inconvénients."

The preceding remarks belong, however, to the earlier period of the revolutionary war, "quand les charretiers n'étaient point militaires, ... et ne marchaient souvent que par la crainte des cannoniers qui les y forçaient." But necessity soon produced that improvement which was requisite to accomplish rapid movements; "il était necessaire d'abord qu'ils fussent soldats, et ensuite qu'ils eussent une instruction toute spéciale. On forme des bataillons du train de l'artillerie."

At a later period, alluding to the fifty guns brought, as we have seen, by Davoust against the enemy's left at Wagram, and the hundred and ten pieces, chiefly of foot artillery, particularly employed by Napoleon to force the centre, the author in question says :—

Moveable artillery decides a battle.

"Nous devons observer l'influence toujours croissante que la mobilité de cette arme lui fait acquérir sur le sort des batailles. Dans celles-ci, dont nous avons dû, pour abréger, supprimer beaucoup de faits importants, la victoire semble toujours appartenir à celui qui sait réunir le plus de pièces sur le point qu'il veut attaquer ou défendre. Le fait qui doit attirer toute notre attention, puisqu'il a décidé le sort de la bataille, c'est la formation de la grande batterie au centre; c'est surtout sa manœuvre dans l'attaque des villages. Que cette batterie arrête l'ennemi, et donne à toute l'armée le temps d'executer un mouvement nécessaire, c'est ce que nous avons souvent vu faire à l'artillerie; mais quand, plus tard, nous la voyons avancer, s'ouvrir à droite et à gauche, livrer ainsi passage à la colonne qui la suit, et chasser seule l'ennemi de deux villages, contre lesquels tant d'efforts avaient echoué, nous n'hésitirons pas à reconnaître un progrès dans l'art d'employer l'artillerie, et à lui attribuer le premier rôle dans cette bataille, dont la perte aurait compromis l'armée entière.

Foot and horse artillery equally available.

"Dans ces grands mouvements de l'artillerie, l'artillerie à pied rivalise avec l'artillerie à cheval, l'histoire ne fait plus de distinction entre ces deux parties de la même arme." *

---

* Histoire et Tactique des Trois Armes, et plus particulièrement de l'Artillerie de Campagne, par Ild. Favé, Capitaine d'Artillerie, pp. 219, 247, 248.

Still as no change, however perfect in the abstract, can be carried out beneficially without the requisite number of officers being effective, it may be advisable to say a few words respecting officers who are employed in military or civil appointments.

Continued employment of artillery officers.

It appears* that the late Sir John Macleod, K.C.H., was appointed to the staff of the artillery in 1783, being then a captain, and that he died a full general, holding the situation of deputy adjutant-general, Royal Artillery, in 1833; so that he had been fifty years without doing regimental duty.

In 1806 Captain Charles Baynes was appointed assistant to the deputy adjutant-general, and continued so employed till his decease in 1818.†

These instances, it is true, relate to the past time; but the present does not differ materially in this respect, since there are at this moment four officers who have not been on regimental duty for periods respectively of thirty-eight, thirty-three, fifteen, and ten years.

The necessity of continuing to the country the services of those who have performed their non-military duties so satisfactorily is not for a moment questioned by the writer; he merely ventures to show the necessity of seconding officers, instead of

* Kane's List, pp. 10-32.

† Ibid., p. 22.

allowing their duties to fall on others for such a length of time.

The benefits of reorganization

With respect to any *possible* or *supposed* benefit to the writer and his cotemporaries which might be the result of the proposed reconstruction, it must be self-evident (to use the words of the answer given to the Committee), that after forty-seven years' service, "It is impossible to benefit those of *my standing*. Nothing can bring back the time we have lost, owing to the defective construction of the regiment which has hitherto prevailed."* The benefits, therefore, belong to the future and to others, since the advantages of any beneficial change that may be devised will in reality only tell when the senior officers shall have passed away, and have been replaced by others in the prime of life, and in such proportion as will secure an outlet by promotion for their juniors.

an only be prospective.

Working of the proposed organization.

The author hopes he has shown that this might be accomplished without expense, simply by having a lieutenant-colonel to every two captains. To effect such a change it is only necessary to add a few gunners, with a lieutenant, and a lieutenant-colonel to each of the present companies, which would then be sufficient for eight guns, and would become a field-officer's instead of a captain's com-

* No. 5775. Report on Army and Ordnance Expenditure, printed by order of House of Commons, 12th July, 1849.

mand as heretofore. The advantages of this change would be, that instead of six guns and five officers being allotted to a captain, eight guns and six officers would have a field-officer. This would undoubtedly be a great advantage compared with the present arrangement of companies under a regimental captain. But something more than additional lieutenant-colonels is absolutely necessary to place the artillery officer on a footing with his cotemporaries of the line, and the following schedules give the details of the higher organization that is now submitted to the public, as the probable means of ameliorating the state of the Royal Regiment of Artillery.

SCHEDULE No. 1.

HORS

EXPENSE OF THE PRESENT ESTABLISHMENT.

No.	No.		£.	s.	d.		£.	s.	d.
1	..	Colonel Commandant, at . . .	1,095	0	0	per annum	1,095	0	0
2	..	Colonels, Second	1	12	4	per diem	1,180	3	4
1	..	Major-General on Colonel's Pay .	1	12	4	,,	590	1	8
3	..	Lieut.-Colonels	1	7	1	,,	1,482	16	3
1	..	Lieut.-Colonel	1	2	11	,,	418	4	7
7	..	Captains	0	17	2	,,	2,193	0	10
7	..	Second Captains	0	16	1	,,	2,054	12	11
1	..	Adjutant	0	17	9	,,	323	18	9
1	..	Quartermaster	0	10	10	,,	197	14	2
21	..	First Lieutenants	0	9	10	,,	3,768	12	6
..	1	Sergeant-Major	0	4	4¼	,,	79	9	3¼
..	1	Quartermaster-Sergeant . . .	0	3	10¼	,,	70	6	9¼
..	1	Staff-Sergeant	0	3	9¼	,,	68	16	4½
..	1	Trumpet-Major	0	2	10	,,	52	0	0
..	1	Farrier and Smith	0	3	10¾	,,	71	1	11¾
..	1	Carriage-Smith	0	3	10¾	,,	71	1	11¾
..	3	Ditto	0	3	4¾	,,	185	18	5¼
..	1	Collar-Maker	0	3	4¾	,,	61	19	5¾
..	14	Troop Staff-Sergeants	0	3	9¼	,,	963	8	11½
..	21	Sergeants	0	2	10	,,	1,085	17	6
..	21	Corporals	0	2	4	,,	894	5	0
..	14	Bombadiers	0	2	2	,,	553	11	8
..	354	Gunners	0	1	5¼	,,	9,286	19	4½
..	7	Trumpeters	0	2	1¼	,,	268	16	1¾
..	136	Drivers	0	1	3¼	,,	3,154	4	2
..	7	Trumpeters	0	1	5¼	,,	183	12	9¾
..	7	Farriers and Carriage-Smiths .	0	3	4¾	,,	433	16	4¼
..	9	Shoeing and Carriage-Smiths .	0	2	3¼	,,	372	19	8¼
..	7	Collar-Makers	0	2	0¾	,,	263	9	8¼
..	7	Wheelers	0	2	0¾	,,	263	9	8¼
..	7	Troops' Allowance	36	10	0	per annum	280	0	0
45	621					£	31,969	10	3¾

SCHEDULE No. 1.

BRIGADE.

EXPENSE OF THE PROPOSED ESTABLISHMENT.

No.	No.		£.	s.	d.		£.	s.	d.
1	..	Colonel Commandant, at . . .	1,095	0	0	per annum	1,095	0	0
1	..	Colonel, Second	1	12	4	per diem	590	1	8
1	..	Major-General on Colonel's Pay .	1	12	4	,,	590	1	8
2	.	Lieut.-Colonels.	1	7	1	,	988	10	10
1	..	Lieut.-Colonel, Second Class . .	1	2	11	,,	418	4	7
7	..	Captains, 1 being a Rocket Troop	0	17	2	,,	2,193	0	10
1	..	Adjutant	0	19	9	,,	323	18	9
1	..	Quartermaster	0	10	10	,,	197	12	2
14	..	Lieutenants	0	9	10	,,	2,512	8	4
..	1	Sergeant-Major	0	4	4¼	,,	79	9	3¼
..	1	Quartermaster-Sergeant, at . .	0	3	10½	,,	70	6	9½
..	1	Staff-Sergeant	0	3	9¼	,,	68	16	4¼
..	1	Trumpet-Major	0	2	10	,,	52	0	0
..	1	Farrier and Smith	0	3	10¾	,,	71	1	11¾
..	1	Carriage-Smith	0	3	10¾	,,	71	1	11¾
..	3	Ditto	0	3	4¾	,,	185	18	5¼
..	1	Collar-Maker	0	3	4¾	,,	61	19	5¾
..	7	Troop Staff-Sergeants	0	3	9¼	,,	481	14	5¾
..	14	Sergeants	0	2	10	,,	723	18	4
..	21	Corporals	0	2	4	,,	894	15	0
..	14	Bombadiers	0	2	2	,,	553	11	8
..	7	Trumpeters	0	2	1¼	,,	268	16	1¾
..	336	Gunners	0	1	5¼	,,	8,814	15	0
..	120	Drivers	0	1	3¼	,,	2,783	2	6
..	7	Farriers and Carriage-Smiths .	0	3	4¾	,,	433	16	4¼
..	7	Collar-Makers	0	2	0¾	,,	263	9	8¼
..	7	Shoeing and Carriage-Smiths .	0	2	3¾	,,	290	1	11¾
..	7	Wheelers	0	2	0¾	,,	263	9	8¼
29	557						25,596	15	11¼
		Decrease					6,372	14	4½
						£	31,969	10	3¾

SCHEDULE No. 2.

REGIMENT OF

EXPENSE OF THE PRESENT ESTABLISHMENT.

No.	No.		£.	s.	d.		£.	s.	d
12	..	Colonels Commandant, at . . .	1,003	0	0	per annum	12,036	0	0
12	..	Colonels on Major-Generals' Pay, at	1	6	3	per diem	5,748	15	0
24	..	Colonels, Second	1	6	3	,,	11,497	10	0
36	..	Lieut.-Colonels	0	18	1	,,	11,880	15	0
12	..	Lieut.-Colonels	0	16	11	,,	3,704	15	0
96	..	Captains	0	12	2	,,	21,316	0	0
96	..	Second Captains	0	11	1	,,	19,418	0	0
..	..	Allowance for Captains of Companies	36	10	0	,,	3,894	0	0
192	..	First Lieutenants	0	6	10	,,	23,944	0	0
96	..	Second Lieutenants	0	5	7	,,	9,782	0	0
12	..	Adjutants	0	12	9	,,	2,792	5	0
12	..	Quartermasters	0	7	10	,,	1,715	10	0
..	12	Sergeant-Majors	0	4	1¼	,,	898	16	3
..	12	Quartermaster Sergeants . .	0	3	7¼	,,	789	6	3
..	12	Farriers and Carriage-Smiths .	0	3	2¾	,,	707	3	9
..	12	Shoeing and Carriage-Smiths .	0	2	1¼	,,	460	16	3
..	12	Collar-Makers	0	1	10¾	,,	414	13	9
..	12	Wheelers	0	1	10¾	,,	414	13	9
..	96	Company-Sergeants	0	3	2	,,	5,548	0	0
..	288	Sergeants	0	2	8	,,	14,016	0	0
..	384	Corporals	0	2	2	,,	15,184	0	0
..	384	Bombadiers	0	2	0	,,	14,016	0	0
..	192	Trumpeters	0	1	3¼	,,	4,453	0	0
..	9,000	Gunners and Drivers	0	1	3¼	,,	208,734	7	6
..	..	Repair of Arms and Poundage for 96 Companies . . .					3,894	0	0
..	..	Postage and Stationery for ditto					672	0	0
600	10,416	Total . . . . £					394,038	8	0

	Officers.	Staff and Non-Commissioned Officers.		Gunners and Drivers.		Non-Commissioned Officers and Gunners.	£.	s.	d
Horse Brigade . .	45	131	+	490	=	621	31,969	19	3¾
Battalions . . .	600	1,416	+	9,000	=	10,416	394,038	8	0
Present Cost of both Services . . .							426,007	18	3¾

Decrease of Officers, 16.
Increased Annual Expense caused by the Battalion organization, 859*l*. 13*s*. 3*d*.

SCHEDULE No. 2.

ARTILLERY.

EXPENSE OF THE PROPOSED ESTABLISHMENT.

No.	No.		£.	s.	d.		£.	s.	d.
12	..	Colonels Commandant, at . .	1,003	0	0	per annum	12,036	0	0
12	..	Colonels Commandant on Major-Generals' Pay	1	6	3	per diem	5,748	15	0
24	..	Colonels, Second	1	6	3	,,	11,497	10	0
48	..	Lieut.-Colonels	0	18	1	,,	15,841	0	0
24	..	Lieut.-Colonels, Second Class .	0	16	11	,,	7,409	10	0
144	..	Captains	0	12	2	,,	31,974	0	0
..	..	Allowance for ditto, at . . .	36	10	0	each	5,256	0	0
144	..	First Lieutenants.	0	6	10	per diem	17,958	0	0
144	..	First Lieutenants, at reduced pay	0	5	7	,,	14,673	10	0
24	..	Adjutants	0	12	9	,,	5,584	10	0
24	..	Quartermasters	0	7	10	,,	3,431	0	0
..	24	Sergeant-Majors	0	4	1¼	,,	1,797	12	6
..	24	Quartermaster-Sergeants . .	0	3	7¼	,,	1,578	12	6
..	3	Farriers and Carriage Smiths .	0	3	2¾	,,	176	15	11¼
..	3	Shoeing and Carriage Smiths .	0	2	1¼	,,	115	4	0¾
..	3	Collar-Makers	0	1	10¾	,,	103	13	5¼
..	3	Wheelers	0	1	10¾	,,	103	13	5¼
..	144	Company-Sergeants	0	3	2	,,	8,322	0	0
..	288	Sergeants	0	2	8	,,	14,016	0	0
..	432	Corporals	0	2	2	,,	17,082	0	0
..	432	Bombadiers	0	2	0	,,	15,768	0	0
..	144	Trumpeters	0	1	3¼	,,	3,339	15	0
..	8,916	Gunners and Drivers	0	1	3¾	,,	206,786	3	9
..	..	Stationery and Postage for 144 Companies	4	13	4	,,	672	0	0
600	10,416	Total . . . . £					401,270	15	7¼
							394,038	8	0
		Excess . . . . £					7,232	7	7¼

	Officers.		Staff and Non-Commissioned		Gunners and Drivers.		Non-Commissioned Officers and Gunners.	£.	s.	d.
Proposed Horse Brigade	29		101	+	456	=	557	25,596	15	11¼
Proposed Battalions .	600		1,500	+	8,916	=	10,416	401,270	15	7¼
Proposed Cost of both Services . .								426,867	11	6¾

SCHEDULE No. 3.

*Outline showing the Companies at present serving at the Out-Stations in Great Britain and Ireland, and those of the Battalions to be formed from the 48 Companies supposed to remain for these duties after strengthening the Foreign Garrisons.**

Companies.	Stations.	Companies Proposed.		New Companies.		Battalions.	Proposed Allotment.
16	Woolwich	16	to become	24	=	4	4
11	Ireland	10	,,	15	=	2½	2½†
3	Portsmouth	4	,,	6	=	1	1
4	Devonport and Plymouth	4	,,	6	=	1	1
3	Sheerness and Chatham.	4	,,	6	=	1	1
2	Dover	2	,,	3	=	0½	1 (Dover, Jersey, Guernsey, &c.)
1 1 1	Jersey . . . . . Guernsey . . . . Alderney	2	,,	3	=	0½	
2	Scotland	2	,,	3	=	0½	1 (Scotland and N. of England.)
2	Hull, Weedon, and Harwich	2	,,	3	=	0½	
1 1	Manchester . . . . Liverpool and Chester .	2	,,	3	=	0½	0½
		48	,,	72		12	12

* N. B.—The number at Woolwich is supposed to be diminished for this purpose, and also one company in Ireland.

† The other half at Manchester, Liverpool, and Chester.

SCHEDULE No. 4.

Outline of a Change of Construction in the Royal Regiment of Artillery, showing the present Stations of the 48 *Companies supposed to be stationed Abroad, as well as those of the new Battalions into which they are to be formed. Four Companies at or near the Stations are to compose a Battalion of Six Companies, Field Officers being attached according to Tours of Service; the Formation and Promotions are to be simultaneous; Officers of all Ranks to remain where they are, as far as there are Vacancies; and the remainder removed to the several Battalions in which there are vacancies, according to their position on the Rolster.*

Companies at present.	Stations.	Companies Proposed.		New Companies.		Small Battalions.	Proposed Allotment of small Battalions.
5	Gibraltar	8	to become	12	=	2	2
3	Malta	4	,,	6	=	1	1
3	Ionian Islands	4	,,	6	=	1	1
7	Canada	8	,,	12	=	2	2
2	Halifax	2	,,				
1	New Brunswick	1	,,	6	=	1	1
1	Newfoundland	1	,,				
3	Jamaica and Bahamas						
4	Barbadoes	8	,,	12	=	2	2
1½	Bermuda						
1	St. Helena	1	,,	6	=	1	1
2	Cape	3					
2	Mauritius						
2	Ceylon	8	,,	12	=	2	2
1	China						
0½	Australia and New Zealand						
		48		72		12	12

A skeleton company, recruiting at home, would be advisable for at least some of the Battalions stationed abroad, under the charge of some of the Officers who are absent from sickness or other causes.

SCHEDULE No. 5.

Outline showing the Details of a Battery or Company of Garrison Artillery, and also of the Battalions in Peace time, whether of Garrison or Field Artillery; viz. in the latter case with a Brigade of Field Batteries on the Peace Establishment.

THE BATTERY OR COMPANY.

Colonels-en-Second.	Lieut.-Colonels.	Captains.	Adjutant.	Quartermaster and Acting Paymaster.	Lieutenants.	Surgeons or Assistant Surgeon.	Sergeant-Major.	Quartermaster Sergeant.	Staff-Sergeants.	Sergeants.	Corporals.	Bombardiers.	Gunners and Drivers, including Batmen.	Trumpeters.	Farriers.	Carriage Smiths.	Shoeing Smiths.	Collar Makers.	Wheelers.	Horses	Officers.	Total Non-Commissioned Officers and Men.	Total Officers, Non-Commissioned Officers and Men.
..	..	1	..	..	2	..	..	..	1	2	3	3	58	1	1	1	..	1	1	50	3	72	75

THE BATTALION.

Colonels-en-Second.	Lieut.-Colonels.	Captains.	Adjutant.	Quartermaster and Acting Paymaster.	Lieutenants.	Surgeons or Assistant Surgeon.	Sergeant-Major.	Quartermaster Sergeant.	Staff-Sergeants.	Sergeants.	Corporals.	Bombardiers.	Gunners and Drivers, including Batmen.	Trumpeters.	Farriers.	Carriage Smiths.	Shoeing Smiths.	Collar Makers.	Wheelers.	Horses	Officers.	Total Non-Commissioned Officers and Men.	Total Officers, Non-Commissioned Officers and Men.
1	2	4	1	1	8	1	1	1	4	8	12	12	232	4	4	4	8	8	4	200	17	302	319*
..	1	2	..	..	4	2	..	..	2	4	6	6	116	2	2	2	4	4	2	100	7	150	157
1	3	6	1	1	12	3	1	1	2	12	18	18	348	6	6	6	12	12	6	300	24	452	476

Total 24 Officers, and 452 Non-commissioned Officers, Trumpeters and Artificers for the Battalion in Peace time, or 600 Officers and 10,416 Non-commissioned Officers, Artificers, Gunners, &c.; and 600 Officers, 14,487 Non-commissioned Officers, Gunners, &c. on the full establishments, exclusive of the Horse Artillery.

On taking the field for a Campaign, the 6-pounder, with two spare ammunition waggons, is supposed to have from 86 to 90 horses, and the 9-pounder Battery also with two spare waggons, 100 or 104 horses; the spare horses and ammunition being, according to the proposed arrangement, with the Reserve.

* Reserve usually not mounted on Home Service.

SCHEDULE No. 6.

Outline of a farther Change, dividing the Twenty-four Battalions into Eighteen Battalions of Heavy or Garrison Artillery, and Six Battalions of Field Artillery or Brigades of Batteries, with the allotment of each Service. The Officers and Men for the Field, being in the first instance by selection.

In time of Peace, the Eighteen Battalions of Heavy or Garrison Artillery, and the Six Field Batteries or Brigades of Batteries, might be thus distributed:—

Heavy or Garrison Artillery.			Field Artillery.		Batteries.	Guns.
Woolwich	2	Battalions.	1½	Battalions or Brigades of Batteries, &c. viz.	6 3	or 24 Horsed. or 12 without Horses,
Ireland	1	,,	1½	,, ,,	6 3	or 24 Horsed. or 12 without Horses.
Portsmouth, Devonport, and other Home Stations	4	,,	2	,, ,,	8 4	or 32 Horsed. or 16 without Horses.
Canada	1	,,	1	,, ,,	4 2	or 16 Horsed. or 8 without Horses.
Halifax and Newfoundland	1	,,	..		..	..
Gibraltar	2	,,	..		..	..
Malta	1	,,	..		..	..
Ionian Islands	1	,,	..		..	..
Jamaica, Barbadoes, Bermuda, &c.	2	,,	..		..	..
St. Helena and the Cape	1	,,	..		..	..
Ceylon, Mauritius, China, Australia, &c.	2	,,	..		..	..
	18		6	Total	36 Batteries	or 96 Guns Horsed, and 48 in reserve without Horses.

Stations of the Horse Brigade.

1 Troop at Newcastle.
1 Troop at Leeds.
2 Troops at Woolwich.
1 Troop (Rocket) at Woolwich.
2 Troops in Ireland.

The Second Captains who are not required for the Seven Troops on the reduced scale, will be absorbed by the promotion in the higher Ranks.

SCHEDULE No. 7.

Comparative Expense of an Artillery Force of

PRESENT ORGANIZATION.				£.	s.	d.
Brought forward—Expense of 45 Officers and 621 Non-commissioned Officers and Gunners of the Horse Brigade.				31,969	10	3¾
Brought forward—Expense of 600 Officers and 10,416 Non-commissioned Officers and Gunners				394,038	8	0
Additional for an Increase of Three Battalions.						
3 Colonels Commandant				3,009	0	0
6 ditto ditto Second, at	£1	6	3	2,874	7	6
9 Lieut.-Colonels	0	18	1	2,970	3	9
3 ditto ditto	0	16	11	926	3	9
24 Captains	0	12	2	5,329	0	0
24 Second Captains	0	11	1	4,854	10	0
3 Adjutants	0	12	9	698	1	3
3 Quartermasters	0	7	10	428	17	6
48 Lieutenants	0	6	10	5,986	0	0
24 Second ditto	0	5	10	2,445	10	0
3 Sergeant-Majors	0	4	1¼	224	14	0¾
3 Quartermaster Sergeants	0	3	7¼	197	6	6¾
24 Company Sergeants	0	3	2	1,387	0	0
72 Sergeants	0	2	8	3,504	0	0
96 Corporals	0	2	2	3,796	0	0
96 Bombadiers	0	2	0	3,504	0	0
719 Gunners	0	1	3¼	86,346	9	0¼
24 Companies, allowance for repair of Arms and Poundage				973	10	0
24 Companies allowance for Postage and Stationery .				168	0	0
Total for 792 Officers and 15,044 Non-commissioned Officers, Gunners, &c. . . . . . £				555,537	16	3¾

In 1814, according to Kane's List,* there were (exclusive of the Officers and Men of the Driver Corps) 678 Officers of all Ranks, in the Horse Brigade and Battalions, and 16,902 Non-commissioned Officers, Gunners, &c., or 1 Officer to 25 Men.

* Page 82.

SCHEDULE No. 7.

15,044 *Non-commissioned Officers and Gunners.*

PROPOSED ORGANIZATION.	£.	s.	d.
Brought forward—Expense as proposed for 29 Officers and 557 Non-commissioned Officers and Gunners of the H.B.	25,596	15	11½
Brought forward—Expense as proposed for 600 Officers and 10,416 Non-commissioned Officers and Gunners of 24 Battalions on reduced Establishment	401,270	15	7½
For the supposed increase to the Full Establishment.			
144 Additional Sergeants	7,008	0	0
144 ditto Corporals	5,694	0	0
3739 Gunners, Drivers, Trumpeters, Artificers, &c. . .	86,717	10	8¾
Total for 629 Officers and 15,044 Non-commissioned Officers, Gunners, &c.	527,287	2	3¾
Decrease on 15,044 Men	29,250	14	0
£	555,537	16	3¾

The proposed organization of 629 Officers of all Ranks for the Horse Brigade and Battalions, and 15,044 Non-commissioned Officers and Gunners, &c., would give 1 Officer to 24 Men.

INDEX.

LONDON: PRINTED BY WILLIAM CLOWES AND SONS, STAMFORD-STREET.